Green Energy and Technology

More information about this series at http://www.springer.com/series/8059

Annoula Paschalidou · Michael Tsatiris
Kyriaki Kitikidou · Christina Papadopoulou

Using Energy Crops for Biofuels or Food: The Choice

Springer

Annoula Paschalidou
Department of Forestry and Management of the Environment and Natural Resources
Democritus University of Thrace
Orestiada
Greece

Kyriaki Kitikidou
Department of Forestry and Management of the Environment and Natural Resources
Democritus University of Thrace
Orestiada
Greece

Michael Tsatiris
Department of Forestry and Management of the Environment and Natural Resources
Democritus University of Thrace
Orestiada
Greece

Christina Papadopoulou
Aristotle University of Thessaloniki
Thessaloniki
Greece

ISSN 1865-3529 ISSN 1865-3537 (electronic)
Green Energy and Technology
ISBN 978-3-030-09706-6 ISBN 978-3-319-63943-7 (eBook)
https://doi.org/10.1007/978-3-319-63943-7

Printed on acid-free paper

This Springer imprint is published by the registered company Springer International Publishing AG part of Springer Nature
The registered company address is: Gewerbestrasse 11, 6330 Cham, Switzerland

Contents

Concepts and Abbreviations

AFOLU	Agriculture, forestry, and other land use
ARPA-E	Advanced Research Projects Agency-Energy
AUA	Agricultural University of Athens
Bioenergy	Energy from biomass (does not include, e.g., wind, solar, or hydroelectric power)
Biofuel	Solid, liquid, or gaseous fuel produced from biomass. For example, wood chips, ethanol, and biogas
Biomass	The biodegradable fraction of products, waste, and residues of biological origin from agriculture (including vegetable and animal matter), forestry and related industries, including fisheries and aquaculture, as well as the biodegradable fraction of industrial and municipal waste
BNEF	Bloomberg New Energy Finance
Bnl	Billion liters
BTU	British thermal unit
CAP	Common Agricultural Policy
CFS	Committee on World Food Security
CO	Carbon monoxide
CO_2	Carbon dioxide
CRES	Center for Renewable Energy Sources and Saving
EC	European Commission
EEA	European Economic Area
Energy crops	The crops cultivated in the field for bioenergy purposes
ES	Expected shortfall
EU	European Union
EV	Electron volt
FAO	Food and Agriculture Organization
FITs	Feed-in tariffs
FQD	Fuel Quality Directive
GDP	Gross domestic product

GHG	Greenhouse gases
GM	Genetic modification
Gm3	Billion cubic meters
GMOs	Genetically modified organisms
HLPE	High-Level Panel of Experts
IEA	International Energy Agency
IEO	International Energy Outlook
IFPRI	International Food Policy Research Institute
ILUC	Indirect land use
INDC	Intended Nationally Determined Contributions
KW	Kilowatt
LCA	Life cycle analysis
MMT	Million metric tons
MS	Member State
MW	Megawatt
MWh	Megawatt hours
OECD	Organization for Economic Co-operation and Development
OPEC	Organization of the Petroleum Exporting Countries
OPEKEPE	Greek Payment Authority of Common Agricultural Policy
OPTIMA	OPTImization of perennial grasses for biomass production in Mediterranean Area
PETRO	Plants Engineered to Replace Oil
PV	Photovoltaics
R&D	Research and development
RED	Renewable Energy Directive
RES	Renewable energy sources
SDG	Sustainable development goal
SWOT	Strengths, weaknesses, opportunities, and threats
TOE	Tonnes of oil equivalent
UAA	Utilized agricultural area
UN	United Nations
US	United States
WEO	World Energy Outlook

List of Figures

List of Tables

Abstract

The challenge of feeding a constantly increasing population and covering the insatiable energy demand of contemporary economies is a difficult one for human race and may prove fatal for the environment and the planet. Although biofuel production from energy crops attempts to address the energy problem, it leads to a serious dilemma: "energy crops for biofuel production or for food?" This paper aims to examine these two options through SWOT analysis. As well as identifying the positive and negative points of each option considering a large variety of environmental, social, economic, and sustainability aspects, this paper also suggests an array of strategies and alternative choices for sustainable, green energy supply and adequate nutrition for the earth's growing population.

Keywords Energy crops · Food · Biofuel · Bioenergy · SWOT analysis

Introduction

Today, mankind is facing three basic threats: hunger, the lack of energy, and the deterioration of the environment. The challenge is to eliminate all three simultaneously, since any one of them can singularly pose a grave threat to humanity and civilization [1]. Covering humanity's nutrition needs presents major difficulties which become even more when trying to anticipate future global population growth [2]. According to Food and Agriculture Organization (FAO) data, in 2050 more than nine billion people will have to share the planet's limited resources. In order to cover these needs, food production will have to rise by 50% over present levels [3]. This estimate can even reach 60% or 70% according to other forecasts [4, 5]. The global demand for food, fuel, and fiber will place an increasing pressure on natural capital, land, and water resources. The question whether these resources will be adequate may become academic since they are already in peril due to climate change and must not be taken for granted. Moreover, along with earth's climate, ecosystems and habitats are among the first to face threats from human presence and consumption [4, 6]. Consequently, it will be extremely difficult to feed the world's growing population in a way that it does not harm the soil, water, and biodiversity and generally complying with environmental goals. This will be feasible only if energy systems and agricultural produce change dramatically [3].

On the other hand, global energy demand is increasing even faster than population growth. Fossil fuels, apart from being a depletable energy source, entail catastrophic results for the environment so seeking alternative energy sources has been a priority for many countries for the past decades. Renewable energy sources like solar, wind, and bioenergy have been gaining momentum and are a constant subject of research and development globally [7]. Bioenergy, which was the main source of power and heat prior to the industrial revolution, currently provides roughly 10% of global supplies and accounts for roughly 80% of the energy derived from renewable sources [6]. Many multinational corporations from powerful countries have been involved in the production of biodiesel and bioethanol [8]. In 2010, bioenergy was the source of approximately 7.5% of energy used in the EU. Biofuels are an important component of the renewable energy share in EU's energy plan, particularly in the transport sector, in an effort to mitigate climate change [9].

Countries like Brazil and USA are leading the biofuel production sector. The development of the biomass energy industry has also contributed to the improvement of the rural economy [10].

This rapid expansion of bioenergy production has not come without its costs, especially since almost all of the commercially available biofuels are produced from either starch- or sugar-rich crops (for bioethanol), or oilseeds (for biodiesel) [11, 12]. It has resulted in a significant shift of arable land and agricultural products to biofuels instead of food or feed. The displacement of agricultural production has been discussed extensively and is generally called the indirect land-use change (ILUC) effect [13]. This land-use change has been blamed for increasing food prices, intensive agrochemical use, stress on water and other resources, deforestation, land degradation, and biodiversity risks [6, 14]. Deforestation especially is regarded as the greatest threat to the environment as forests are a major storage of carbon and the most biologically diverse terrestrial ecosystems [15, 16]. Moreover, large-scale bioenergy production may further intensify existing pressure on land and resources with huge implications for ecosystems and biodiversity [17, 18].

Most importantly, bioenergy has raised the ethical issue of agricultural production being used for energy when millions of people in the world are starving [8]. The most controversial aspect of biomass production so far, aptly termed as "the food versus fuel debate," is the fact that it competes with food and feed supply and threatens food security, especially since first-generation biofuels are produced from soy, palm, and rapeseed oils or starch and sugar crops like maize, wheat, or sugarcane, which are all valuable food resources [3].

The belief that increased food production and energy provision should be practiced in an environmentally friendly way, which does not deplete natural resources, has resulted in the evaluation of biomass production according to sustainability principles [19]. In order to continue biomass use, as a means to reduce fossil fuel use, without food and environmental risks, research has focused on second- and next-generation biofuels such as lignocellulosic feedstocks, perennial grasses, and algae among others, with significant results [20, 21].

This paper provides a comprehensive review on global nutritional need, the energy problem, and bioenergy, especially biofuel production from energy crops. The impact on food production, land requirements, and the environment are also discussed. The objective of this paper is to investigate the two options of the choice "Using energy crops for biofuels or food" through SWOT analysis, in order to identify the strengths, weaknesses, opportunities, and threats regarding each one of the two options. By taking into consideration a large variety of environmental, social, economic, and sustainability aspects, strategies are formulated and suggested, as well as alternative courses of action, in order to make informed strategic choices than benefit the people, the planet, and the environment, now and in the future.

Chapter 1 addresses the increasing nutrition needs and the growing energy demands of world population and how these are to be covered by the specific arable land and the finite resources of the planet. It also explores the need for new renewable energy sources and the efficient use of the existing ones.

Chapter 2 describes the concept of bioenergy, as a renewable energy source from energy crops, providing biomass production data for Europe and the world, together with the features of biodiesel- and bioethanol-producing crops, both annual and perennial.

Chapter 3 presents the main biofuel categories and their world production to date. It also presents the major environmental, geopolitical, financial, and social reasons for the increased biofuel production and consumption.

Chapter 4 examines biofuels from a sustainability perspective and stresses the necessity of planning and developing energy systems that observe sustainability principles.

Chapter 5 introduces the two options regarding the cultivation and the use of energy plants, i.e., growing energy crops for biofuels or for food and provides details about the origins of the debate and the basic concerns and controversies of the two choices.

Chapter 6 presents the methodology tool that was used for investigating the above-mentioned two options, namely SWOT analysis, the steps followed as well as the environmental, social, and financial aspects taken into consideration.

Chapter 7 discusses the two options, i.e., "growing energy crops for biofuel or for food" and identifies the opportunities and threats based on the external analysis, and their existing strengths and weaknesses as evaluated by the internal analysis. The results are then presented in a matrix.

Chapter 8 lists potential alternative strategies based on combinations of the four strategic factors (WT, ST, WO, and SO) of the two options.

Chapter 9 presents a concise account of the basic results and conclusions of the analysis as well as alternative strategies for judicious energy planning and resource management.

References

1. Escobar JC, Lora ES, Venturini OJ, Yáñez EE, Castillo EF, Almazan O (2008) Biofuels: environment, technology and food security. Renew Sustain Energy Rev. http://www.elsevier.com/locate/rser. Accessed 12 Nov 2016 [Elsevier]
2. Pruyt E, De Sitter G (2008) Food or energy? Is that the question? In: Proceedings of the 26th international conference of the system dynamics society. http://www.systemdynamics.org/conferences/2008/proceed/papers/PRUYT372.pdf. Accessed 18 Feb 2017
3. Spiess WEL, Does biofuel production threaten food security, IUFoST food security task force. https://doi.org/10.1007/978-1-4614-7906-2_31. http://www.iufost.org/iufostftp/Does%20Bio-fuel%20Production%20Threaten%20Food%20Security.pdf, http://link.springer.com/chapter/10.1007%2F978-1-4614-7906-2_31. Accessed 03 Dec 2016
4. The New Climate Economy, Report (2016) The sustainable infrastructure imperative: financing for better growth and development. http://newclimateeconomy.report/2016/. Accessed 28 Jan 2017
5. FAO (2011) World livestock 2011—livestock in food security. FAO, Rome. http://www.fao.org/docrep/014/i2373e/i2373e.pdf. Accessed 30 Jan 2017
6. IEA Bioenergy (2009) A sustainable and reliable energy source. Main report. International Energy Agency, Paris. http://www.ieabioenergy.com/wp-content/uploads/2013/10/MAIN-

REPORT-Bioenergy-a-sustainable-and-reliable-energy-source.-A-review-of-status-and-prospects.pdf. Accessed 27 Dec 2016
7. Popp J, Lakner Z, Harangi-Rákos M, Fári M (2014) The effect of bioenergy expansion: food, energy, and environment. Renew Sustain Energy Rev 32:559–578. http://www.sciencedirect.com/science/article/pii/S1364032114000677. Accessed 20 Dec 2016
8. Tolios Y (2012) Tribute of Marxist thought the environmental crisis, vol. 4. January–March 2012, pp 173–85. http://marxistikiskepsi.gr/. Food crisis: society, biodiversity, environment. http://www.aristerovima.gr/blog.php?id=3018. Accessed 12 Nov 2016
9. EC. Renewable energy, national renewable energy action plans. European Commission, 2010. http://ec.europa.eu/energy/renewables/action_plan_en.htm. Accessed 09 Feb 2017
10. Zhang Z, Liu C, Zhang W et al (2015) Achieving food security and high production of bioenergy crops through intercropping with efficient resource use in China. Front Agric Sci Eng 2(2):134–143. http://engineering.cae.cn/fase/EN/10.15302/J-FASE-2015069#AbstractTab. Accessed 27 Dec 2016
11. Crutzen JP, Mosier RA, Smith AK, Winiwarter W (2008) N_2O release from agro-biofuel production negates global warming reduction by replacing fossil fuels. Atmos Chem Phys 8:389–395. Available at: http://www.atmos-chem-phys.net/8/389/2008/acp-8-389-2008.pdf. Accessed 09 Feb 2017
12. Searchinger T, Heimlich R, Houghton AR, Dong F, Elobeid A, Fabiosa J et al (2008) Use of US croplands for biofuels increases greenhouse gases through emissions from land-use change. Science 319:1238–1242. http://whrc.org/wp-content/uploads/2015/09/SearchingeretalScience08.pdf. Accessed 09 Feb 2017
13. Fargione J, Hill J, Tilman D, Polasky S, Hawthorne P (2008) Land clearing and the biofuel carbon debt. Science 319(5867):1235–1238. https://doi.org/10.1126/science.1152747. https://www.ncbi.nlm.nih.gov/pubmed/18258862. Accessed 09 Feb 2017 [Epub 2008, Feb 7]
14. Gumpenberger M, Vohland K, Heyder U, Poulter B, Macey K, Rammig A et al (2010) Predicting pan-tropical climate change induced forest stock gains and losses—implications for REDD. Environ Res Lett 5(1). IOP Publishing Ltd. http://iopscience.iop.org/article/10.1088/1748-9326/5/1/014013/meta. Accessed 09 Feb 2017
15. Hassan R, Scholes R, Ash N (eds) (2005) Ecosystems and human well-being: current state and trends findings of the condition and trends working group. Island Press, Washington, DC, pp 77–122. http://www.millenniumassessment.org/documents/document.766.aspx.pdf. Accessed 09 Feb 2017
16. Laurance WF (2007) Have we overstated the tropical biodiversity crisis? Trends Ecol Evol 22:65–70. https://www.ncbi.nlm.nih.gov/pubmed/17011069. Accessed 09 Feb 2017
17. Erb KH, Gaube V, Krausmann F, Plutzar C, Bondeau A, Haberl H (2007) A comprehensive global 5 min resolution land-use data set for the year 2000 consistent with national census data. J Land Use Sci 2:191–224. http://www.tandfonline.com/doi/abs/10.1080/17474230701622981. Accessed 09 Feb 2017
18. Fingerman K, Torn M, O'Hare M, Kammen D (2010) Accounting for the water impacts of ethanol production. Environ Res Lett 5:014020. http://iopscience.iop.org/article/10.1088/1748-9326/5/1/014020/meta. Accessed 09 Feb 2017
19. Ros JPM, Overmars KP, Stehfest E, Prins AG, Notenboom J, van Oorschot M (2010) Identifying the indirect effects of bio-energy production. Netherlands Environmental Assessment Agency, Bilthoven, The Netherlands. http://www.pbl.nl/sites/default/files/cms/publicaties/500143003.pdf. Accessed 09 Feb 2017
20. Schmer RM, Vogel PK, Mitchell BR, Perrin KR (2008) Net energy of cellulosic ethanol from switchgrass. Proc Nat Acad Sci USA 105:464–469. http://www.pnas.org/content/105/2/464.full. Accessed 09 Feb 2017
21. Melillo MJ, Reilly MJ, Kicklighter WD, Gurgel CA, Cronin WT, Paltsev S et al (2009) Indirect emissions from biofuels: how important. Science 326:1397–1399. http://science.sciencemag.org/content/326/5958/1397. Accessed 09 Feb 2017

Chapter 1
Global Nutritional Need—The Energy Problem

1.1 Global Nutritional Need

The perceived limits to producing food for a growing global population have been a source of debate and preoccupations for ages. Population increased to 6.9 billion in 2010, up from 2.5 billion in 1950 and 3.7 billion in 1970 and according to the United Nations (UN) could reach 9.15 billion in 2050 [1].

According to Food and Agriculture Organization (FAO), food needs of the future will vary dramatically from region to region. Solutions will have to be tailored to each situation, which will require new approaches to production, trade and development planning at national, regional, and inter-regional levels [2]:

- Industrialized countries needs are expected to remain constant or even to decline [2].
- The needs of Latin America, and countries with maize dominated nutritional patterns, are expected to double by 2050 [2].
- An almost triple increase is predicted for the Arab countries and West Asia, i.e., the wheat zone. This will result in major differences in those countries that are financially able to import and those that are not [2].
- Countries with a rice-based diet, mainly Asian, will more than double their needs. Since land is at a premium in this region, they will have to increase productivity and import to meet the additional demand [2].
- The needs of sub-Saharan African countries will present a dramatic increase, considerably greater than those of the rest of the world: In countries with millet- and sorghum-based diet, needs may be multiplied almost fivefold, and where the diet is dominated by roots and tubers by more than sevenfold. However, to cover this increased demand, Africa might not be able to utilize its major land reserves, because it lacks the resources to develop them [2].

Consequently, Asia and sub-Saharan Africa are projected to be areas with potential food security issues by the middle of the next century [2].

A. Paschalidou et al., *Using Energy Crops for Biofuels or Food: the Choice*,
Green Energy and Technology, https://doi.org/10.1007/978-3-319-63943-7_1

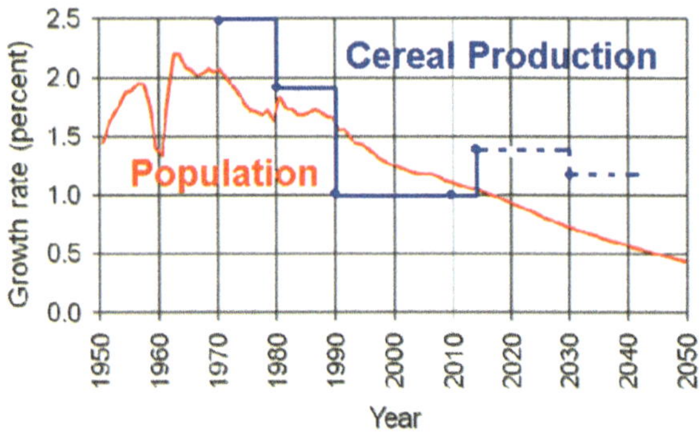

Fig. 1.1 Population and cereal production growth rates (*Source* [14])

International statistical data predict much slower world population growth rates in the following years, while cereal production growth rates will stabilize at a level of 1.2%. However, it is estimated that by 2030 the average daily caloric intake will be 3050 kcal/capita*day compared with today 2800 kcal/capita*day. In other words, the growing world population will absorb the increased production of cereals by increasing the average daily caloric intake in the following years [3].

By 2030, the developing countries will produce only 86% of the required cereal supply. Consequently, they will have to increase net imports from currently 103 to 265 MMT by 2030. This is expected to lead to higher prices which will mainly affect underprivileged portions of the population in developing countries (Fig. 1.1) [3].

Despite the fact that, on a global level, the total food production can adequately feed the population, not all people are well fed. The obvious cause for that is poverty, which is, in many low-income countries, the main reason behind the underdeveloped agriculture, and the limited access to imported food [1].

The agricultural policy in the last decade in Greece as well as the new Common Agricultural Policy (CAP) of the European Union (EU) has been leading agriculture to complete disassociation from the intensified food crisis. The fact that subsidies are not essentially linked to actual production, combined with low prices for producers and increased production cost, contributed to the diminishing of rural income both in volume and in value [4]. It is important to point out that 30 years ago Greece had a positive trade balance, while it is currently importing 40% of the food consumed. The local varieties have been replaced by few genetically improved varieties (hybrids), almost completely [5]. These developments in the agricultural sector have dramatically reduced the nutritional self-sufficiency of the country, resulting in the increase of imports and the trade deficit and the volume of "food addiction" [4].

1.2 The Energy Problem

The disparity between the perpetually increasing energy demands and the diminishing energy reserves, together with the detrimental environmental effect of power consumption, has come to be known as the energy problem.

The fact that energy reserves are not unlimited and that a country's national independence is closely linked to its energy policy became painfully apparent in 1973, when the first energy crisis broke out and oil prices increased dramatically overnight. For the first time, governments put forward energy saving programmes and considered alternative sources instead of fossil fuels [6]. A new course of action both on a local and international level seemed exigent to address the problem, taking into consideration pressing matters such as global warming, the exploding energy demand of emerging economies, and the need for secure energy supply, in light of the fluctuation in fuel prices [6].

Most countries have taken measures to face the energy problem, and a lot has been done globally over the last 40 years. However, despite the ongoing efforts of a large part of the world, the energy problem is still a grave concern and a top priority in every country's policy.

According to International Energy Outlook 2016 (IEO 2016), energy demands will continue to rise over the next three decades, particularly in countries outside of the Organization for Economic Cooperation and Development (OECD). Non-OECD Asia, including China and India, accounts for more than half of the world's total increase in energy consumption over the 2012–2040 projection period. By 2040, energy use in non-OECD Asia exceeds that of the entire OECD by 40 quadrillion British thermal units (Btu) in the IEO 2016 Reference case (Fig. 1.2) [7].

Moreover, the same Reference case reports a 48% increase in the total world energy consumption by 2040. It is worth noting that while OECD nations will present a 18% energy consumption growth, non-OECD nations' respective increase, driven

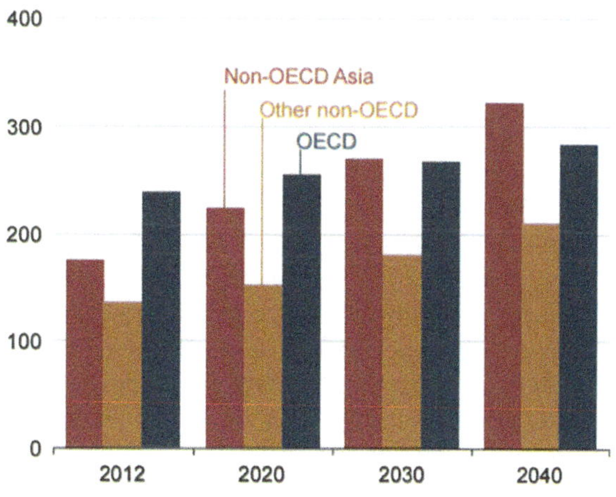

Fig. 1.2 World energy consumption by country grouping, 2012–40 (quadrillion Btu) (*Source* [7])

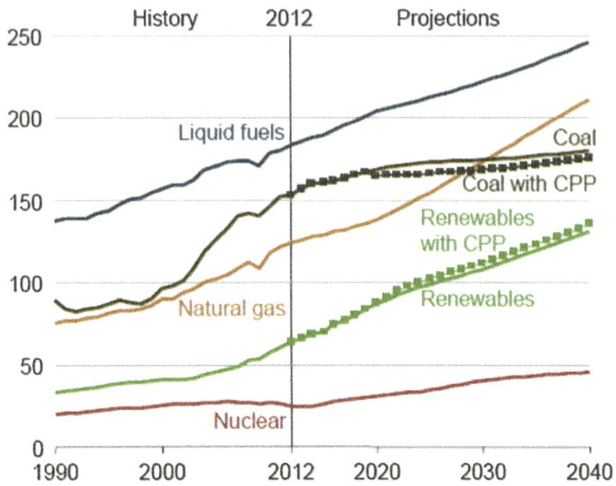

Fig. 1.3 Total world energy consumption by energy source, 1990–2040 (quadrillion Btu) (*Source* [7])

by strong, long-term economic development, will reach 71% by 2040 (Fig. 1.2) [7]. By 2040, almost two-thirds of the world's primary energy will be consumed in the non-OECD economies. It is apparent, therefore, that the major factor that affects energy demand is economic growth—as measured in gross domestic product (GDP). According to the said Reference case, while the world's GDP (expressed in purchasing power parity terms) rises by 3.3%/year from 2012 to 2040, the fastest rates of growth are projected for the emerging, non-OECD countries, where combined GDP increases by 4.2%/year. In OECD countries, GDP grows at a much slower rate of 2.0%/year over the projection as a result of their more mature economies and slow or declining population growth trends (Fig. 1.2) [7].

According to the IEO2016 Reference case, worldwide marketed energy consumption from all fuel sources will increase through 2040. While renewable energy, with an average 2.6%/year growth, will be the fastest-growing energy source, nuclear power with 2.3%/year is close second (Fig. 1.3) [7]. Generally, non-fossil fuel consumption is expected to grow faster than fossil fuels. On the other hand, fossil fuels still account for 78% of energy use in 2040, with natural gas being the fastest-growing fossil fuel in the outlook by 1.9%/year, due to the abundance of natural gas resources. Oil prices account for the decline of liquid fuel consumption 33% in 2012–30% in 2040, although they remain the largest source of world energy. The slowest growing energy source is coal, which rises by 0.6%/year and is surpassed by natural gas by 2030 (Fig. 1.3) [7].

According to European Environment Agency (2009), the primary energy use in EU categorized by fuel can be seen in Fig. 1.4.

Finally, according to the International Energy Outlook 2013, through 2040, new energy sources like clean fossil fuel and renewable energy sources, which will employ the new technologies in the energy sector, will partly replace fossil fuel production (Figs. 1.5 and 1.6) [6, 8].

Since oil formation and accumulation occurred due to specific geological features of certain areas, it is apparent that the world's oil reserves are not equally distributed

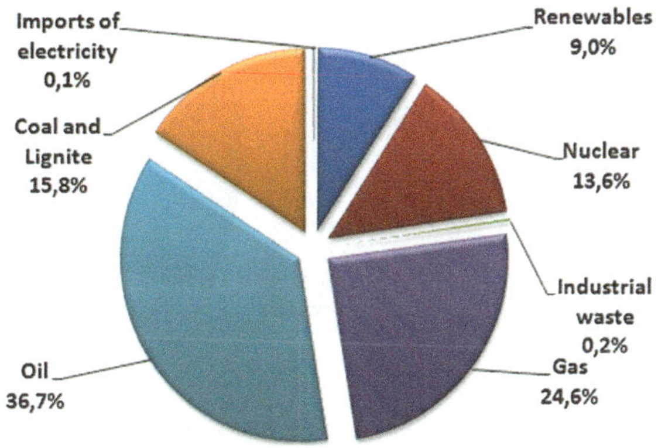

Fig. 1.4 Primary energy use by fuel in EU (2009) (*Source* [15])

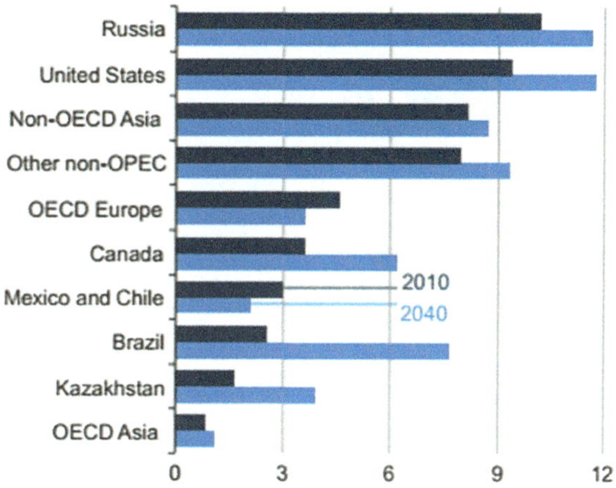

Fig. 1.5 Non-OPEC liquids production by region and country, 2010 and 2040 (million barrels per day) (*Source* [2])

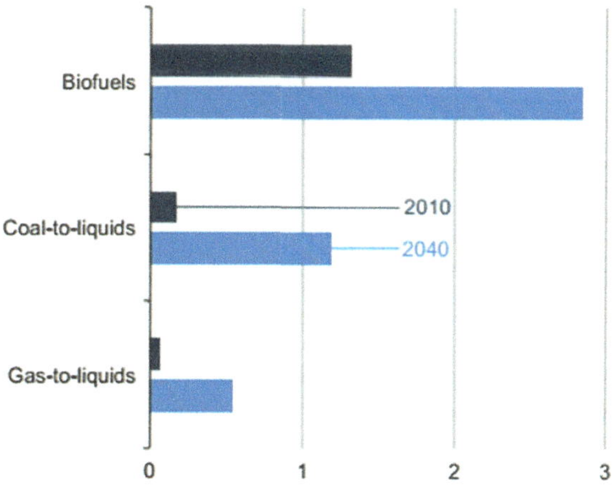

Fig. 1.6 World non-petroleum liquids production by type, 2010 and 2040 (million barrels per day) (*Source* [2])

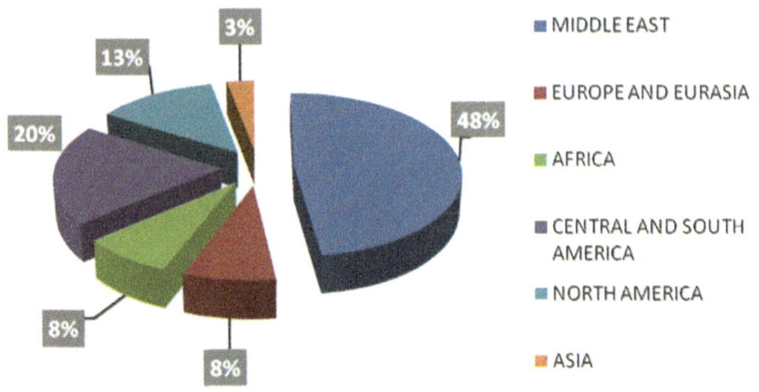

Fig. 1.7 World's oil reserves in 2012 (*Source* [10])

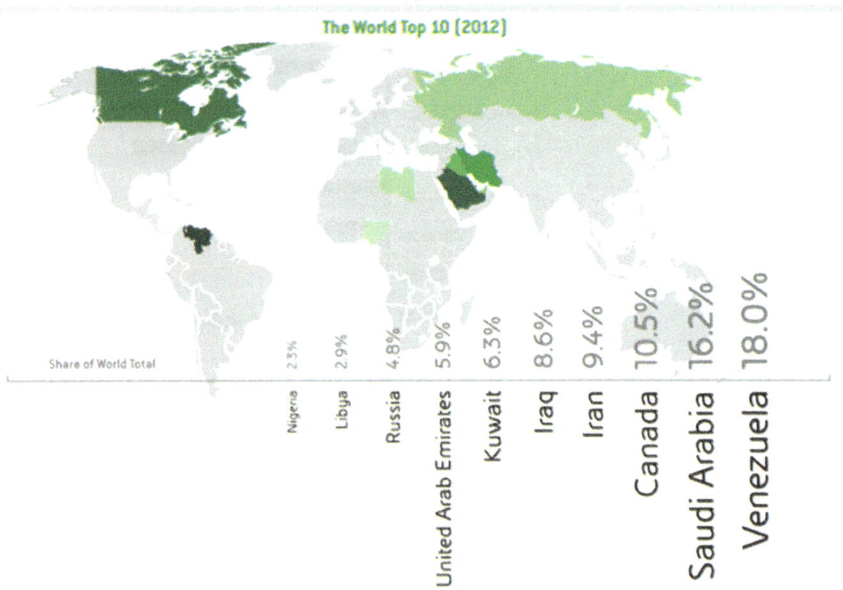

Fig. 1.8 World top ten oil reserve holders in 2012 (*Source* [16])

across the globe [9]. The Middle East concentrates about 48% of the world's reserves, whereas Europe and Eurasia have 8%, Africa 8%, Central and South America 20%, North America 13%, and Asia 3% (Figs. 1.7 and 1.8) [10].

In Greece, the main energy sources are: a) the renewable energy sources and b) the fossil fuels (lignite and, in the future, perhaps the Greek hydrocarbons) [11].

Combination of Energy Sources of Greece in 2010, according to Eurostat, consists of (a) 6% renewable energy sources (RES), (b) 21% coal and lignite, (c) 9% natural gas (imported), (d) 1% electricity imports, (e) 41% oil (Fig. 1.9).

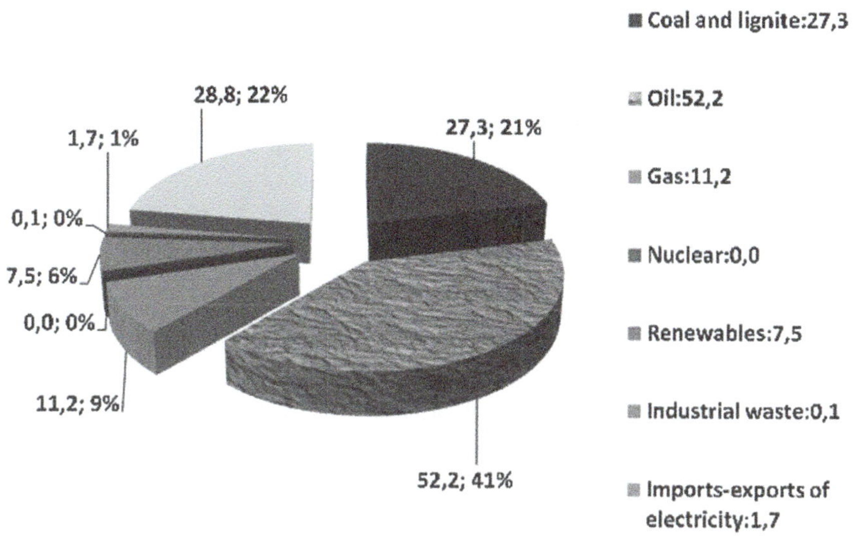

Fig. 1.9 Primary energy consumption by fuel (%) in 2010 (Greece) (*Source* [17])

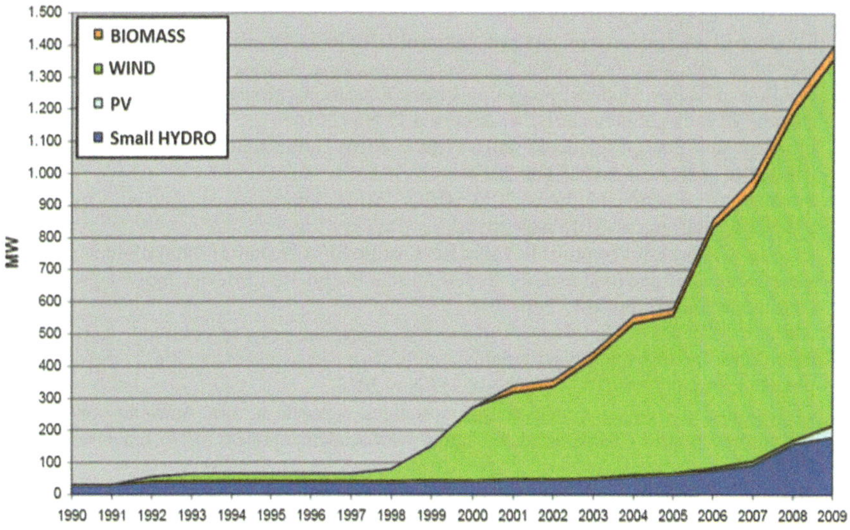

Fig. 1.10 Renewables installed power (2009 data) (*Source* [12])

Over the last years, Greece has attempted to introduce all kinds of renewable sources (RES) with measures such as feed-in tariffs (FITs). In 2011, RES produced energy that increased to percentage of 11.6%, when at 2010, it was 9.2%. Detailed data of installed power of RES can be seen in Fig. 1.10. Today, the RES total installed energy is higher [12].

A substantial part of the Greek energy policy for the past ten years has been the development of renewable energy sources in order to improve the Greek environmental indicators and, in particular, CO_2 emissions. The national target is to achieve 20% contribution of RES energy to the total energy consumption by 2020 [13].

References

1. Alexandratos N, Bruinsma J (2012) World agriculture, towards 2030/2050, the 2012 revision. ESA Working Paper No. 12-03, Agricultural Development Economics Division, Food and Agriculture Organization of the United Nations, 1. http://www.fao.org/fileadmin/templates/esa/Global_perspectives/world_ag_2030_50_2012_rev.pdf. Accessed 12 Nov 2016
2. Food Needs and Population, World Food Summit Food for All, Food and Agriculture Organization of the United Nations, Rome (1996) http://www.fao.org/docrep/x0262e/x0262e23.htm. Accessed 15 Nov 2016
3. Spiess WEL (2013) Does biofuel production threaten food security. IUFoST Food Security Task Force. https://doi.org/10.1007/978-1-4614-7906-2_31. http://www.iufostorg/iufostftp/Does%20Biofuel%20Production%20Threaten%20Food%20Security.pdf, http://link.springer.com/chapter/10.1007%2F978-1-4614-7906-2_31. Accessed 03 Dec 2016
4. Tolios Y (2011) Nutritional crisis. National Consumers, VIOZO, Athens. http://www.biozo.gr/node/584. Accessed 20 Nov 2016
5. Common Place, Food sovereignty and judgment Tritsi Park, Athens (2010). http://koinostopos.espivblogs.net/, http://biotechwatch.gr/diatrofikiautarkeia. Accessed 20 Nov 2016
6. The Role of Energy, Modern Energy Challenges (Educational program for energy). http://www.cie.org.cy/sxoliko.html#menu1-2-4. Accessed 15 Nov 2016
7. International Energy Outlook 2016 (IEO 2016) http://www.eia.gov/outlooks/ieo/pdf/0484(2016).pdf. Accessed 28 Feb 2017
8. International Energy Outlook 2013 (IEO 2013) http://www.eia.gov/outlooks/ieo/pdf/0484(2013).pdf. Accessed 15 Nov 2016
9. Escobar JC, Lora ES, Venturini OJ, Yáñez EE, Castillo EF, Almazan O (2008) Biofuels: environment, technology and food security. Renew Sustain Energy Rev. Elsevier. http://www.elsevier.com/locate/rser. Accessed 12 Nov 2016
10. Ahtsoglou G (2014) The electric car and forecasting the price of crude oil. University of Macedonia, Interdisciplinary postgraduate studies in business administration, 3. https://dspace.lib.uom.gr/handle/2159/16020. Accessed 15 Nov 2016
11. Christoforou L (2012) Greek energy resources. Speech at the Academy of Athens, November 13, 2012 Athens Academy. http://www.academyofathens.gr/Documents/ellinikoi_energeiakoi_poroi.pdf. Accessed 20 Nov 2016
12. Katsivelis P (2013) Energy in Greece. Basic information and figures, power imbalances-alternatives for the energy sector in Greece and its European and global context, Public Conference, Rosa Luxemburg Stiftung Brussels and Athens. http://rosalux.gr/sites/default/files/report_katsivelis_rozalux.pdf. Accessed 12 Nov 2016
13. Greece in the USA, About Greece—Energy and Environment. http://www.mfa.gr/usa/en/about-greece/energy-and-environment/energy.html. Accessed 15 Dec 2016
14. Spiess WEL (2013) Does biofuel production threaten food security. IUFoST Food Security Task Force. http://www.iufost.org/iufostftp/Does%20Biofuel%Production%20Threaten%20Food%20Security.pdf. Accessed 12 Nov 2016
15. European Environment Agency (2009) Primary energy use by fuel in EU (2009). http://www.eea.europa.eu/data-and-maps/indicators/renewable-primary-energy-consumption/renewable-primary-energy-consumption-assessment-7. Accessed 11 Mar 2017

16. O&G World Oil and Gas Review (2013) World oil and gas review 2013. https://www.eni.com/docs/en_IT/enicom/publications-archive/publications/wogr/2013/O-G_2013_WEB.pdf. Accessed 11 Mar 2017
17. Eurostat http://ec.europa.eu/eurostat/search?p_auth=hhdBKuxF&p_p_id=estatsearchportlet_WAR_estatsearchportlet&p_p_lifecycle=1&p_p_state=maximized&p_p_mode=view&_estatsearchportlet_WAR_estatsearchportlet_action=search&text=greece. Accessed 30 Jan 2017

Chapter 2
Bioenergy–Biomass–Energy Crops

2.1 Bioenergy–Biomass

Bioenergy is energy derived from biomass. The most important sources for biomass are derived from agricultural areas, where plants like food and energy crops are cultivated and forest areas. A secondary biomass source is agricultural and municipal waste residues (Fig. 2.1) [1].

Bioenergy is utilized in three core energy applications: electricity, heat, and fuels. Bioenergy, in this case, faces not only competition from fossil energy, but also from other sustainable energy sources, such as hydro energy and wind power, which have a significant share as alternative energy sources. However, electrical power derived from biomass has had a remarkable growth rate during recent years [1]. The use of biomass for energy is estimated at about 14% of global primary energy, with the majority being used in developing countries where biomass covers up to 1/3 of the energy needs [2]. In addition, biomass has other uses such as agrochemicals or pharmaceuticals, with different volume requirements and higher added value according to its end use (Fig. 2.2) [3].

Although biomass can be converted to other usable forms like methane gas, ethanol fuel, and biodiesel fuel [4], it is primarily used for food and feed, followed by bio-based products, biofuels, and bioenergy. Figure 2.3 shows the use of the 10 billion tons of biomass harvested worldwide in 2008. Animal feed predominates with a share of 60%, which will increase even further due to increasing meat consumption [5].

Biomass has the potential to cover half of our supply needs, but at the cost of shifting land away from food production. Currently, approximately 2% of global agricultural land is used for biomass for energy, supplying approximately 10% of the global energy mix (Fig. 2.4). While 2% of the total agricultural land may appear small, the challenge of feeding the world's growing population, combined with water shortages, makes it very difficult to increase this to any great extent [14].

A. Paschalidou et al., *Using Energy Crops for Biofuels or Food: the Choice*,
Green Energy and Technology, https://doi.org/10.1007/978-3-319-63943-7_2

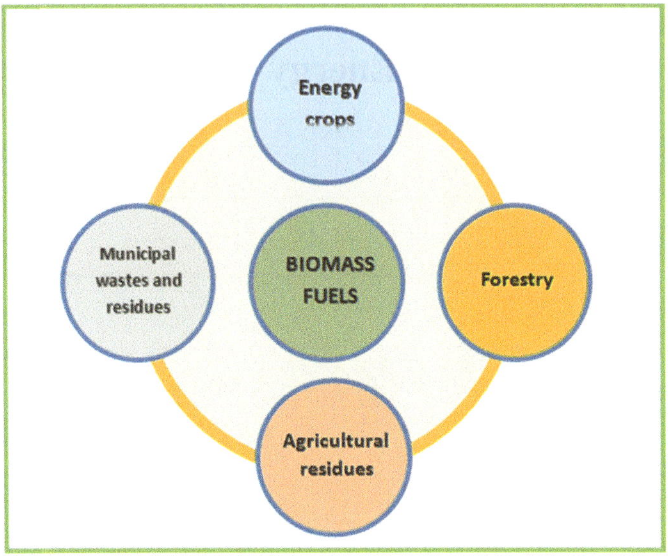

Fig. 2.1 Biofuels (Biomass) sources (*Source* [3])

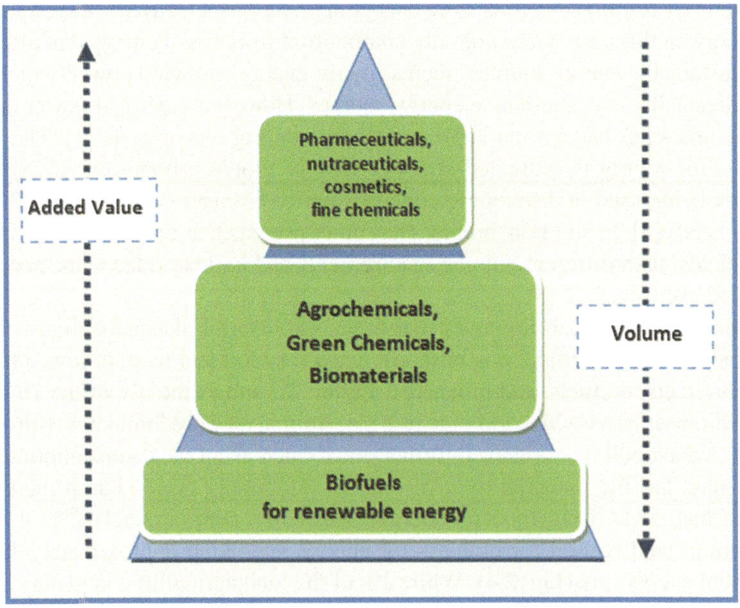

Fig. 2.2 Added value pyramid of biomass uses (*Source* [3])

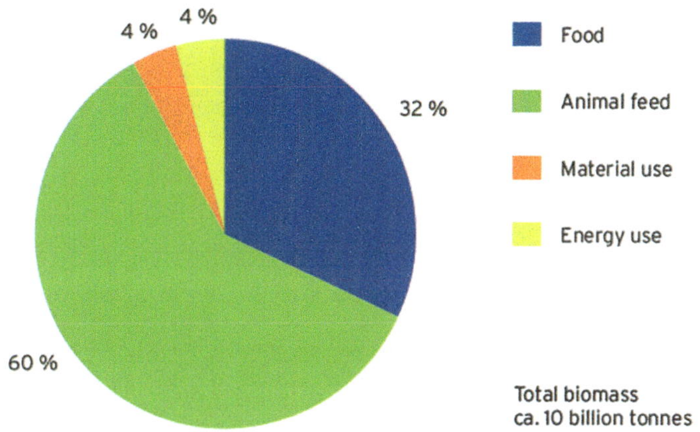

Fig. 2.3 Worldwide allocation of harvested biomass by production target (main product) in 2008 (*Source* [5])

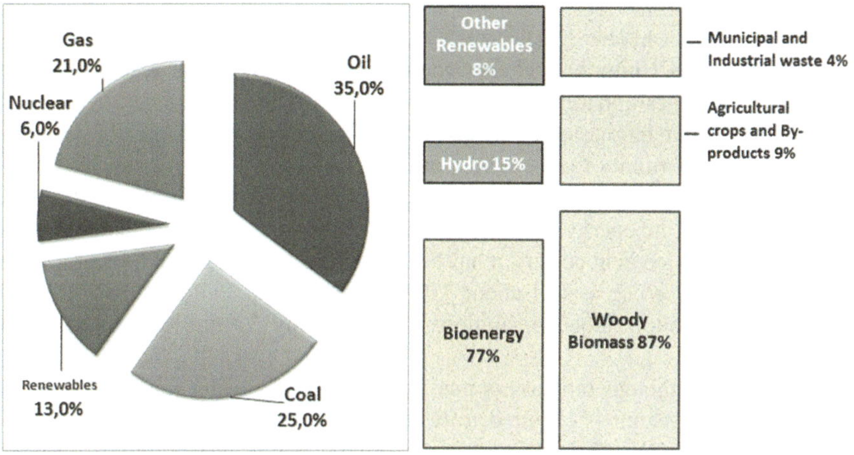

Fig. 2.4 Share of bioenergy in the world primary energy mix (*Source* [38])

Biomass use is a new important trend in Europe acquiring an increasing share as a power source. Its uses in European countries include growing use of solid biomass pellets, use of biomass in building-scale or community-scale combined heat and power plants, and use of biomass for centralized district heating systems [6].

In European Union (EU) between 2000 and 2011, biomass grew by 49 million tones of oil equivalent (toe). Wind and solar only grew by 13 and 6 million toe, respectively. In other words, the absolute biomass growth was 2.5 ($49/(13+16=19)=2.57\approx 2.5$) times greater than in wind and solar, and so far, the majority of new renewable energy since 2000 has come from biomass, not wind or solar (Fig. 2.5) [7].

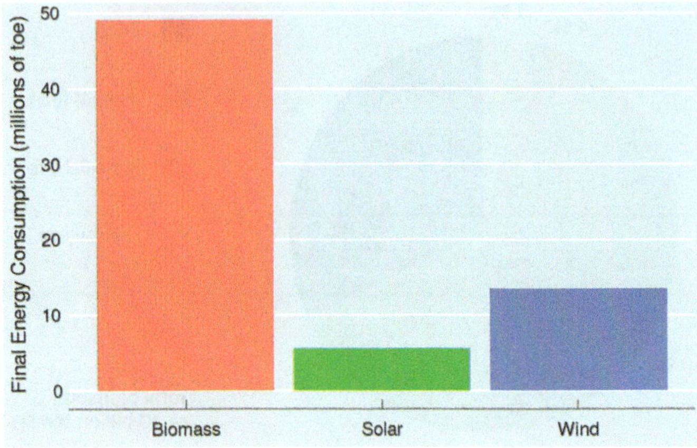

Fig. 2.5 Growth of EU biomass, wind and solar 2000–2011 (*Source* [7])

Biomass-produced energy can reduce oil or gas dependency. The USA, the EU, and recently even China are oil dependents, and in light of the recent oil price increase, it becomes apparent that politicians are favoring domestically generated energy sources like biomass. Moreover, biofuel production is considered by many a development opportunity for the third world [8].

In Greece, according to Centre for Renewable Energy Sources and Saving (CRES), only about 3% of the energy needs is being covered by the use of available biomass [9]. According to a recent census, it has been estimated that the total direct biomass available in Greece consists of about 7500,000 tons of agricultural crop residues (cereal, maize, cotton, tobacco, sunflower, etc.) and by 2.7 million tons of forest wood waste [10].

To sum up, although biomass is one of the most promising renewable energy sources, more research is required to determine whether power generation from biomass is both technically and economically viable, but most importantly to ensure that meets sustainability criteria [4, 11].

2.2 Energy Crops

Energy crops are cultivated or self-sown kinds of plants that produce biomass which can be further used for various energy aims, like heat production, electrical power, and biofuels [12].

To qualify as an energy crop, the cultivation of any crop must meet certain conditions such as:

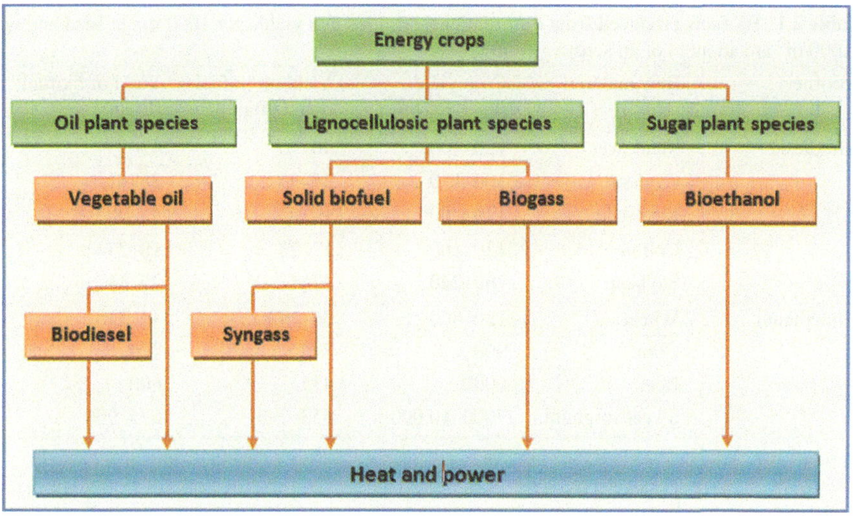

Fig. 2.6 Biomass to bioenergy–energy crops (*Source* [3])

- A positive energy ratio, a positive CO2 ratio, and a positive life cycle analysis [13].
- A reduction of the negative environmental impact of its cultivation such as decrease of soil erosion, improvement of ground fertility, increase of water use efficiency, limitation of fertilizer demands, and more [13].
- Tolerance to soil contamination [3].
- Accumulation of trace elements in aerial biomass [3].
- High yields [3].
- Low production cost [3].
- Environmentally friendly way of cultivation [3].
- Ability to be cultivated in contaminated land with satisfactory yields, giving satisfactory net profit [3].
- Adaptability to the climatic conditions of the contaminated area [3].

Energy crops are divided into two main categories: agricultural and forest energy crops. They can also be categorized according to plant species, cultivation, and development methods and ultimate use (Fig. 2.6).

Energy crops are generally grown for two different markets, i.e., solid biofuels for heat and power generation on one hand, and liquid transport fuels on the other hand (Tables 2.1 and 2.2).

Biodiesel and bioethanol, which are known as first-generation biofuels, dominate the biofuel transport sector today and are mainly produced from annual crops such as rapeseed and cereals. A shift toward second-generation biofuels, produced mainly from lignocellulosic biomass (mostly perennial crops such as miscanthus and salix grown in short rotation), is expected in the future. This shift will greatly affect not only agriculture, but also industry. For example, the development of perennial biomass crops has been stated to be the major driver for the cellulosic ethanol industry [14].

Table 2.1 Biofuels produced from different energy crops and yields per 1000 m^2 in seed and oil (1000 m^2 are about ¼ of an acre)

Biofuel	Raw material	Crop yield (kg/1000 m^2)	Yield of biofuel (kg/1000 m^2)	Yield of biofuel (lt/1000 m^2)
Biodiesel	Sunflower	120–300	40–70	43–75
	Cardoon	100–150	24–36	28–41
	Rapeseed	120–300	40–83	43–90
	Cotton	120–160	17–23	18–25
	Soybean	160–240	27–41	29–44
Bioethanol	Wheat	150–800	36–190	45–240
	Corn	900	213	270
	Beet	6000	475	600
	Sweet sorghum	7000–10,000	553–790	675–900

Source [39]

Table 2.2 Data of energy crops in Greece

Scientific name	Common name	Time duration of the crop	Kind of produced biofuel
Arundo donax	Giant reed	Perennial	Thermal and electrical energy
Cynara cardunculus	Cardoon	Perennial	Thermal and electrical energy
Miscanthus sp.	Miscanthus	Perennial	Bioethanol
Beta vulgaris	Sugarbeet	Annual	Bioethanol
Brassica sp.	Rapeseed	Annual	Biodiesel
Helianthus spp	Sunflower	Annual	Biodiesel
Hibiscus cannabinus	Kenaf	Annual	Thermal and electrical energy
Sorghum bicolor	Sweet sorghum	Annual	Bioethanol
Triticum aestivum	Wheat soft	Annual	Bioethanol
Glycine max	Soybean	Annual	Biodiesel
Gossypium hirsutum	Cotton	Annual	Biodiesel
Panicum virgatum L	Switchgrass	Annual	Thermal and electrical energy—bioethanol
Zea mays	Corn	Annual	Bioethanol

Source [39]

Table 2.3 Energy crops, according to the EU most suitable for southern European countries and especially for Greece

Forest energy crops	Agricultural energy crops	
	Perennial crops	Annual crops
Two types of Eucalyptus (*Eucalyptus globules Labill, Eucalyptus camaldulensis Dehnh*)	Giant reed (*Arundo donax L.*)	Sweet and fiber sorghum (*Sorghum bicolor L.*)
Robina pseudoacacia L.	*Miscanthus x giganteus* GREEF et DEU	*Hibiscus L.*
	Cardoon (*Cynara cardunculus L.*)	*Brassica napus L., Brassica carinata L. Braun*
	Switchgrass (*Panicum virgatum L.*)	

Source [12]

Choosing the most suitable energy crops (rapeseed, sunflower, switchgrass, cardoon, miscanthus, etc.) is a complicated issue, and several key factors have to be taken into account [15] such as existing conversion technologies, energy-type needs, production environments, harvesting periods, storage costs, the general preference for non-irrigated perennial energy crops. The most important factor is the opportunity cost of the land [16].

The energy crops most suitable for southern European countries and especially for Greece, according to the European Union, are divided as shown in Table 2.3 [17].

Bioenergy can be delivered by a variety of crops, cropping systems, and conversion technologies [18]. Two major categories, depending on their life cycle, are annual and perennial energy crops. The positive and negative aspects of annual and perennial energy crops are discussed below.

2.2.1 Perennial Energy Crops

Perennial energy crops are herbaceous plants with a multiyear life cycle, usually about 10–15 years. Such plants include many well-known energy crops like cardoon, miscanthus, switchgrass, giant reed, and reed canary grass [19] (Table 2.3). The study of solar capture by plants reveals C4 perennial energy crops' photosynthesis to be as efficient as photovoltaic devices [20].

Generally, perennial grasses are suitable for biomass production and a great source for second-generation biofuels due to their high-yield potential, the high contents of lignin, cellulose, and hemicelluloses polysaccharides and their positive social and environmental benefits [21].

The perennial lignocellulosic crops combine high biomass productivity with low requirements for inputs, such as nitrogen fertilization, pesticides, tillage operations, and water supply [21, 22]. The use of pesticides is very limited and concentrated primarily to the first two years [23].

Their cultivation presents a number of benefits for the soil. Their extensive root system can potentially bind soil, thus preventing erosion and filtering heavy metal contaminants from wastewaters [24]. Moreover, in comparison with intensive farming, they improve soil structure and reduce soil erosion, particularly when cardoon, miscanthus, or switchgrass replace conventional annual crops [21, 25, 26]. This is also supported by Bical Energy reports [27]. They have been found to reduce nitrogen and phosphorus loss to surface and groundwater compared to arable land uses. Furthermore, they can potentially increase the organic carbon soil content particularly in soils with previously depleted carbon levels [28]. They may also be used to phytoremediate sites contaminated with heavy metals and to treat water waste, farmland drainage, sewage sludge, and landfill leachate [21, 25].

The major advantage of perennial crops is that plowing and planting are not necessary every year. Consequently, while annual irrigation, fertilizers, pesticides, and insecticides are required, the initial cultivation cost can be divided throughout the duration of the crop life. Perennial grasses grow continuously and are generally more able to survive environmental stresses such as drought. Because they require less fertilizers and pesticides, they are ideal for cultivation on sub-optimal, underused, or abandoned land [24].

Rural communities could generate new sources of income, as well as employment opportunities, by growing these perennial bio-factories on marginal lands, without compromising existing crops grown for food [21, 24]. It will also bring about the development of regional economic structures and improve the education, training, and assistance services provided for farmers [21].

In general, perennial crops differ from annual crops by having a viable root system during autumn, winter, and spring when N is potentially leached from the root zone [29]. Loss of N from agriculture reduces biodiversity of natural ecosystems, pollutes drinking water, thereby affecting human health, and contributes to global warming. However, to meet the targets of the EU Water Framework Directive further reductions are necessary, and as a measure toward achieving this, perennial energy crops have been suggested [30].

Perennial energy crops also lead to less greenhouse gas emissions. Compared to annual crops such as cereals, there will be a drop in N_2O emissions equal [23]. In addition, such grasses could potentially help reduce atmospheric greenhouse gasses by acting as carbon sinks [21, 24]. Perennial energy crops will secure carbon storage (sinks) of 1565 tones CO_2 equivalent per hectare [21, 23].

On the other hand, the extended life cycle may seem to farmers as a long-term commitment with unsure prospects, especially when there are no clear market outlines for the end product [19]. Some of them are exclusively planted by rhizomes and/or plants and that makes it more expensive than the seed planting. Consequently, they have lower farmer/public acceptance compared to the annual energy crops [31].

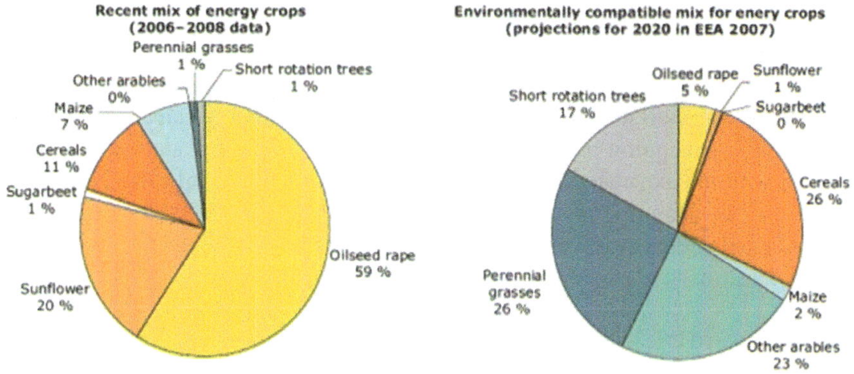

Fig. 2.7 Mix of energy crops, 2006–2008 and European Economic Area (EEA) scenario for environmentally compatible energy cropping in 2020 (*Source* [35])

Moreover, certain agronomic features of perennial grasses such as blossom sterility, high initial installation cost, relative low harvest mechanization, high humidity during harvest, and high ash percentage present considerable weaknesses [32]. Another important drawback is that they use more water than traditional crops and grass land [25].

On the landscape issues, there is concern about "forestification" of the open agricultural landscape, where 4–6-m-high "green walls" of perennial crops blocking the view [23] thus potentially having a significant impact on landscape. If large areas are planted, it could lead to monoculture. Perennial energy crops will benefit some animals connected to more closed landscapes to the disadvantage of birds of the open land, thereby negatively affecting biodiversity [33, 34]. Finally, root depth and hydrological impact may negatively affect archaeology [25].

Europe's current mix of energy crops is not favorable to the environment. To reduce environmental impacts, a broader range of crops is recommended, particularly perennial crops, which are not harvested annually. These could be energy grasses or short rotation willow plantations [35]. The mix of energy crops, (2006–2008) and the European Economic Area (EEA) scenario for environmentally compatible energy cropping in 2020, is presented in Fig. 2.7 [35].

According to 4FCROPS, the most suitable energy crops in terms of agronomic management, climatic adaptability, and potential biomass production in northern Europe are some fast-growing trees and perennial grasses such as poplar, willow, reed canary grass, switchgrass, and miscanthus. In the Mediterranean climate of southern Europe eucalyptus, sweet sorghum, giant reed, and cardoon are promising energy crops (Table 2.3) [31].

The EU-funded "Optimization of Perennial Grasses for Biomass Production" (OPTIMA) project aims to use perennial grasses adapted to the Mediterranean environment as sources for biofuels and novel bio-products. Ideal traits of new varieties include high-yield potential and the ability to grow under water-deficit or high-salt conditions [24].

The ratio of the energy produced from combustion compared to that used to grow, harvest, and transport the crop gives a measure of the net energy gained. The energy ratios of first-generation biofuels are highly variable. Land degradation is a major concern with the expansion of first-generation biofuel crops [36]. These concerns have led to an increased interest in the use of second-generation bioenergy crops and perennial energy crops. The lignocellulose in perennial forage crops is a more energy-dense material than the starch and sugars used from first-generation bioenergy crops [36].

It is estimated that with the progress of the advanced biofuels, the perennial energy crops will become quite an important source of feedstock, especially if the establishment costs are reduced and perennial grasses are domesticated and genetically improved [31].

2.2.2 *Annual Energy Crops*

Annual energy crops have a life cycle of a year or a season and have to be replanted every time. Such crops are fibrous sorghum, kenaf, cannabis, as well as most traditional crops which are suitable for energy production (Tables 2.3 and 2.4) [19]. They are mainly used for the production of first-generation biofuels (bioethanol and biodiesel), while some of them can be used alternatively for food or fiber production [31].

Annual energy crops are widely adopted by farmers because they can be easily integrated with the conventional agricultural systems and the existing rotation systems. Moreover, they can easily be established by seed and can be cultivated with the existing machinery used for other conventional crops with minor adjustments [31].

The similarities with conventional annual crops regarding cultivation and harvest methods ensure cost reduction at low market levels. Furthermore, they can be rotated with other spring crops such as cotton, sugar beet, and corn. Thus, annual energy crops present a feasible, safe choice for farmers, leaving them free from long-term commitment demands, while they can contribute to the biodiversity of an area. Additionally, many of these crops can be used as raw material for the fiber industry or even as feed [19].

However, the disadvantage is that the final fuel price is increased, due to the cost and the energy demands of annual planting, as they have to be replanted every year [19]. Annual energy crops often require more agrochemicals than other bioenergy crops. If crops are cultivated in winter, the risk of soil compaction is high. They are more frequently cultivated than other energy crops, and therefore, the risk of soil erosion may be higher, as well as the risk of nitrogen leaching compared to other bioenergy crops [25].

Table 2.4 Data of annual energy crops in Greece—2013

Region	Kind of energy crop	Area (ha)	Production (kgr)
Eastern Macedonia and Thrace	Rapeseed	4424.20	1,027,650.00
	Sunflower	458,556.10	107,364,759.80
	Soya bean (soybean)	15,849.60	5,316,352.00
Central Macedonia	Rapeseed	34,391.90	7,862,254.00
	Sunflower	180,613.30	47,829,813.40
	Soya bean (soybean)	2971.60	792,834.00
Thessaly	Sunflower	10,484.80	3,213,507.00
	Soya bean (soybean)	35.90	13,130.00
West Macedonia	Rapeseed	656.30	135,350.00
	Sunflower	18,899.10	5,506,579.00
	Soya bean (soybean)	87.10	5820.00
Central Greece	Sunflower	7470.50	2,343,927.00
Peloponnese	Sunflower	309.40	85,980.00
Total of energy crops		734,749.80	1,027,650.00

Source [40]

Today, annual crops greatly contribute to bioenergy worldwide, but yields of these crops are highly dependent on high N inputs into soils. In contrast, perennial grasses and tree crops can achieve higher biomass yields with relatively lower inputs of nitrogen fertilizer and are regarded as most efficient nitrogen users [37].

Crop residues from the production of major annual row crops, such as corn stover and wheat straw, offer a potential bridge between food-based and dedicated energy crop-derived fuels. However, there are several important sustainability issues linked to harvesting residues from annual cropping systems. Traditionally, annual food crops have been grown in rotation with other crops to promote sustainability of the cropping system. It has been recognized for some time that lignocellulose from perennial species, particularly C4 grasses, is better suited than grain or sugar from annuals for efficient renewable energy production in the temperate zone. The recent trend toward increased continuous production of corn to meet increasing demand due to ethanol production may have negative environmental impacts [20].

References

1. Lenk F, Broring S, Herzog P, Leker J (2007) On the usage of agricultural raw materials—energy or food? An assessment from an economics perspective. Biotechnol. J. 2(12):1498. http://www.biotechnology-journal.com. Accessed 15 Nov 2016
2. Bauen A, Kaltschmitt M (2001) Current use and potential of solid biomass in developing countries and their implications for CO_2 emissions. In: Proceedings of the World

Biomass Conference. London: James & James. https://books.google.gr/books?id=I4md2B_nk9UC&pg=PA1354&lpg=PA1354&dq=Bauen+A,+Kaltschmitt+M+(2001)&source=bl&ots=sXaYwa7aBB&sig=KE3QtlfrUHxYTYLvDHQgQtlKdA4&hl=el&sa=X&ved=0ahUKEwjkvNuV7q3SAhULkRQKHT4ECkkQ6AEIITAB#v=onepage&q=Bauen%20A%2C%20Kaltschmitt%20M%20(2001)&f=false. Accessed 22 Nov 2016

3. Papazoglou GE (2016) Use of energy crops for phytoremediation purposes, SPEED, London, UK.http://www.core-community.net/wp-content/uploads/2016/04/Use-of-energy-crops-for-phytoremediation.pdf. Accessed 26Jan 2017
4. Banos R, Manzano-Agugliarob F, Montoyab FG, Gila C, Alcaydeb A, Gomezc J (2011) Optimization methods applied to renewable and sustainable energy: a review. Renew Sustain Energy Rev 15:1753–1766. http://www.sciencedirect.com/science/article/pii/S1364032110004430. Accessed 10 Dec 2016
5. Carus M, Dammer L, Nova-Institut GmbH nova paper #2 on bio-based economy 2013–07 Food or non-food: Which agricultural feedstocks are best for industrial uses? http://www.ibbnetzwerk-gmbh.com/uploads/media/nova_institut_-_food_or_non-food_paper.pdf. Accessed 08 Dec 2016
6. Benini E, Toffolo A (2002) Optimal design of horizontal-axis wind turbines using blade-element theory and evolutionary computation. J Solar Energy Eng 124(4):357–63. http://solarenergyengineering.asmedigitalcollection.asme.org/article.aspx?articleid=1456592. Accessed 08 Dec 2016
7. Wilson R (2014) Biomass: the hidden face of the Energiewende, Energy Post. Retrieved on 2015-12-03 (http://www.energypost.eu/biomass-hidden-face-energiewende/) Accessed 03 Dec 2016
8. Gyulai I (2008) The biomass dilemma. CEEweb for Biodiversity and the National Society of Conservationists—Friends of the Earth Hungary. http://www.mtvsz.hu/dynamic/biomassen.pdf. Accessed 22 Dec 2016
9. Boutetsiou E, Katsoulakos N (2010) Energy Utilization of Forest Biomass: The case of Metsovo, 21. http://www.ntua.gr/MIRC/6th_conference/presentaions/1_main_sessions/5th_session/BOUTETSIOY%20EL%20KATSOYLAKOS%20N.pdf. Accessed 22 Nov 2016
10. Kittas C, Gemtos T, Fountas S, Barzanas T (2007) Biofuels and energy crops. Renewables-their position in the new energy landscape of the country and in the region of Thessaly, Larissa: TEE–ΚΔΘ. http://library.tee.gr/digital/kdth/kdth_3460/kdth_3460_kittas.pdf. Accessed 22 Nov 2016
11. Evans A, Strezov V, Evans TJ (2010) Sustainability considerations for electricity generation from biomass. Renew Sustain Energy Rev 14(5):1419–27. http://www.crcnetbase.com/doi/abs/10.1201/b17093-3. Accessed 08 Dec 2016
12. Boukis I, Vassilakos N, Kontopoulos G, Karellas S (2009) Policy plan for the use of biomass and biofuels in Greece, Part I: Available biomass and methodology. Renew Sustain Energy Rev 13(5):977. www.elsevier.com/locate/rser. Accessed 28 Jan 2015
13. Vyrlas P (2014) Energy crops and protecting soil from erosion, "THALES" Larisa, GCG. http://biomass.agr.uth.gr/attachments/article/122/Virlas.pdf. Accessed 09 Feb 2017
14. Escobar JC, Lora ES, Venturini OJ, Yáñez EE, Castillo EF, Almazan O (2008) Biofuels: environment, technology and food security. Renew Sustain Energy Rev. http://www.elsevier.com/locate/rser. Accessed 12 Nov2016
15. Giannoulis KD, Danalatos NG, Sakellariou M (2011) Switchgrass, cardoon and miscanthus perennial crops as alternatives for solid biofuel production in Central Greece. University of Thessaly, Dept. of Agriculture, Crop Production & Rural Environment, Volos, Greece. https://www.researchgate.net/publiccation/272942951_SWITCHGRASS_CARDOON_AND_MISCANTHUS_PERENNIAL_CROPS_AS_ALTERNATIVES_FOR_SOLID_BIO-UEL_PRODUCTION_IN_CENTRAL_GREECE. Accessed 26 Apr 2016
16. Lychnaras V, Rozakis S (2006) Economic analysis of perennial energy crop production in Greece under the light of the new CAP. Department of Agricultural Economics & Rural Development Agricultural University of Athens, Greece NEW MEDIT N. 3/2006. http://www.iamb.it/share/img_new_medit_articoli/73_29lychnaras.pdf. Accessed 15 Dec 2016

17. Alexandratos N, Bruinsma J (2012) World agriculture, towards 2030/2050, the 2012 revision. ESA Working Paper No 12-03, Agricultural Development Economics Division, Food and Agriculture Organization of the United Nations, 1. http://www.fao.org/fileadmin/templates/esa/Global_perspectives/world_ag_2030_50_2012_rev.pdf. Accessed 12 Nov 2016
18. Karp A, Richter GM (2011) Meeting the challenge of food and energy security. J Exp Bot 62:3263–3271. http://jxb.oxfordjournals.org/content/62/10/3263. Accessed 10 Dec 2016
19. The Bioenergy System Planners Handbook—BISYPLAN- web-based handbook (2012). http://bisyplan.bioenarea.eu/. Accessed 08 Dec 2016
20. Heaton E, Flavell R, Mascia P, Thomas S, Dohleman F, Long S (2008) Herbaceous energy crop development: recent progress and future prospects. Curr Opin Biotechnol 19:202–209. http://linkinghub.elsevier.com/retrieve/pii/S0958166908000542?via=sd. Accessed 15 Dec 2016
21. Cosentino L S, Scordia D (2011) Optimisation of Perennial Grasses for Biomass Production in Mediterranean Area, Optima (289642), Call: FP7 KBBE-2011-5. http://www.fibrafp7.net/portals/0/OPTIMA.pdf. Accessed 12 Dec2016
22. Kohler B, Pelton M, Katchova A, The pros and cons of growing switchgrass Kentucky. University of Kentucky Department of Agricultural Economics. http://www2.ca.uky.edu/cmspubsclass/tinymce/jscripts/tiny_mce/plugins/filemanager/files/adreum/biofuels/Switchgrass%20SWOT.pdf. Accessed 12 Dec 2016
23. The Danish Ministry of Food, Agriculture and Fisheries (2010) Perennial energy crops Green Energy, The Agricultural Contribution. http://en.mfvm.dk/fileadmin/user_upload/ENGLISH_FVM.DK/Themes/Bioenergy/Perennial_energy_crops.pdf. Accessed 12 Dec 2016
24. European Union (2016) Perennial grasses better for biomass. Project reference: 289642 Funded under: FP7-KBBE, Country: Italy. http://cordis.europa.eu/result/rcn/153956_en.html. Accessed 12 Dec 2016
25. SNIFFER ER05 (2010) Impacts of biomass and bioenergy crops on landscape, land use and the wider environment in Northern Ireland and Scotland. SNIFFER First Floor, Greenside House 25 Greenside Place EDINBURGH EH1 3AA UK, Macaulay Scientific Consulting Ltd Craigiebuckler ABERDEEN AB15 8QH UK. http://www.sniffer.org.uk/files/5013/4183/8003/ER05_Final_Project_Report_v3.1.pdf. Accessed 15 Dec 2016
26. Jankauskas B, Jankauskiene G (2003) Erosion-preventive crop rotations for landscape ecological stability in upland regions of Lithuania. Agricult Ecosyst Environ 95:129–42. http://www.sciencedirect.com/science/article/pii/S0167880902001007. Accessed 12 Dec 2016
27. Bical Energy, Miscanthus environmental profile, Bical Industrial Crops Ltd. (2005) p 10. https://www.yumpu.com/en/document/iew/42909839/miscanthus-environmental-profile-enagri. Accessed 15 Dec 2016
28. F.A.O. (Food and Agriculture Organization), Biofuel development should not com promise food security, says CFS (2013), Rome. http://www.fao.org/news/story/en/item/202208/icode/. Accessed 30 Nov 2016
29. Ahtsoglou G (2014) The electric car and forecasting the price of crude oil. University of Macedonia, Interdisciplinary postgraduate studies in business administration, 3. https://dspace.lib.uom.gr/handle/2159/16020. Accessed 15 Nov 2016
30. Pugesgaard S, Schelde K, Larsen SU, Laerke PE, Jorkensen U (2015) Comparing annual and perennial crops for bioenergy production–influence on nitrate leaching and energy balance. GC Bioenergy 7, 1136–1149. Technology and Food Innovation, Agro Food Park 15, Aarhus N, DK-8200, Denmark Department of Agroecology, Aarhus University. https://doi.org/10.1111/gcbb.12215. http://onlinelibrary.wiley.com/doi/10.1111/gcbb.12215/full. Accessed 10 Dec 2016
31. Alexopoulou E, Gemtos T, Fernando AL, Papatheohari Y, Cosentino SL, Comparison between annual and perennial energy crops, CRES, 19th Km Marathonos Avenue, 19009, Pikermi Attikis, GREECE, 2 University of Thessaly, 3 FCT UNL, 4 AUA, 5 UNICT. http://tarmek.org/bildiriler/Comparison_between_Annual_and_Perennial_Energy_Crops.pdf. Accessed 15 Dec 2016
32. SEE (2012) Energy Efficiency and Renewable Energy Sources—Support of Energy Policy at Local Level, Ener Supply, Manual Renewable Energy. http://ener-supply.eu/downloads/ENER_handbook_en.pdf. Accessed 15 Dec 2016

33. Liu TT, McConkey BG, Ma ZY, Liu ZG, Li X, Cheng LL (2011) Strengths, weaknesses, opportunities and threats analysis of bioenergy production on marginal land. Energy Procedia 5:2381–82. Published by Elsevier Ltd. Selection and peer-review under responsibility of RIUDS. https://www.researchgate.net/profile/Tingting_Liu27/publication/241122933_Strengths_Weaknessness_Opportunities_and_Threats_Analysis_of_Bioenergy_Production_on_Marginal_Land/links/555bfead08aec5ac22329c7d.pdf. Accessed 15 Nov 2016
34. Carneiro P, Ferreira P (2012) The economic, environmental and strategic value of biomass. Renew Energy 44(0):18. Department of Production and Systems, University of Minho, Campus Azurem, 4800-058 Guimaraes, Portugal, 2012 Elsevier Ltd. http://www.energyplan.eu/wp-content/uploads/2013/09/Renewable-energy-scenarios-in-the-Portuguese-electricity-system-ICEE-Conference-2013.pdf. Accessed 15 Nov 2016
35. Producing Eco Friendly Bioenergy, Eco Investor August 2013, Industry Focus. http://www.ecoinvestor.com.au/Articles/Producing-Eco-Friendly-Bioenergy.htm. Accessed 03 Dec 2016
36. Littlejohn C, Curran T, Rainer W, Wratten S (2015) Farmland, food, and bioenergy crops need not compete for land. Solutions 6(3):34–48. http://www.thesolutionsjournal.com/node/237359. Accessed 12 Dec 2016
37. Weih M (2009) Perennial energy crops: growth and management. In: Verheye WH (ed) Crop and soil sciences, Encyclopedia of Life Support Systems (EOLSS), Developed under the Auspices of the UNESCO. Eolss Publishers, Oxford, UK. http://www.eolss.net/Sample-Chapters/C10/E1-05A-28-00.pdf. Accessed 10 Dec 2016
38. Ward S (2010) Food versus Fuel The Great Ethical Dilemma. School of Biosystems Engineering, University College Dublin. https://www.ucd.ie/foodandhealth/non-cms-content/2050/2050-pres-SWard.pdf. Accessed 27 Dec 2016
39. Eleftheriadis I, Possibilities cultivation of energy crops in the Greek area. CRES-CRES, Biomass Department. http://www.lignite.gr/events/eleftheriadis.pdf. Accessed 11 Mar 2017
40. Greek Payment Authority of Common Agricultural Policy (OPEKEPE-ΟΠΕΚΕΠΕ). http://www.opekepe.gr/. Accessed 11 Mar 2017

Chapter 3
Biofuels

Biomass-derived fuels, such as solid biomass, liquid fuels, and various biogases fall under the definition of biofuels [1]. They are generally classified as conventional (first generation) biofuels and advanced biofuels (second, third, and fourth generations). The most common types of the first-generation biofuels, which are made from food crops rich in sugar or starch or vegetable oil, are bioalcohols (ethanol) and biodiesel. Second-generation biofuels (cellulosic ethanol) are produced from residual non-food parts of current crops, as well as other crops that are not used for food or industrial purposes. Third-generation biofuels are obtained from algae. Biofuels created from processes other than the above mentioned are referred to as fourth-generation biofuels. They are highly experimental, and their technologies include decomposition of biofuels at high temperatures, artificial photosynthesis reactions, known as solar-to-fuel, and genetically modified organisms to secrete hydrocarbons [1].

Second and further generation biofuels are still at an experimental level. Although relevant research is supported by both the government and private R&D funds, second-generation biofuel conversion has not yet been fully commercialized and is not expected before 2020. However, technological advancements could lower production costs and accelerate investment and development, thus enabling successful commercialization before that [1, 2].

According to the WEO Scenario, USA and Brazil will continue to be the major biofuel producers closely, followed by Europe. Interestingly, developing countries like Colombia, China, and India will also contribute significantly to the total world production (Fig. 3.1) [3].

Bioethanol is by far the most common biofuel in use worldwide [4]. While bioethanol mainly is produced from the fermentation of corn (*Zea mays*), sugarcane (*Saccharum* spp.), or other starch or sugar-rich crops, it can also be produced from cellulosic materials, including grasses, trees, agricultural residues, or municipal solid wastes, following more complicated processes [5]. Vegetable oil, used frying oil, or animal fat can be used to produce biodiesel through a transesterification process in which oil molecules react with an alcohol and a catalyst to form fatty acid methyl esters and glycerol [5]. Although grains and sugarcane produced ethanol and

A. Paschalidou et al., *Using Energy Crops for Biofuels or Food: the Choice*,
Green Energy and Technology, https://doi.org/10.1007/978-3-319-63943-7_3

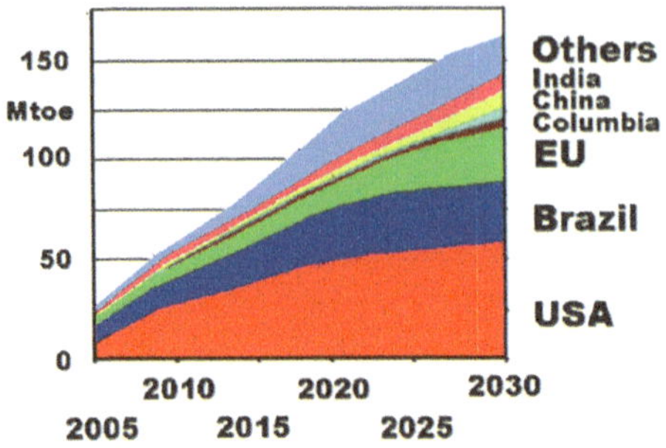

Fig. 3.1 World biofuel production by country [ITP9IEA-WEO 20080 Reference Scenario] [Total: ~5% of total world fuel production] (*Source* [3])

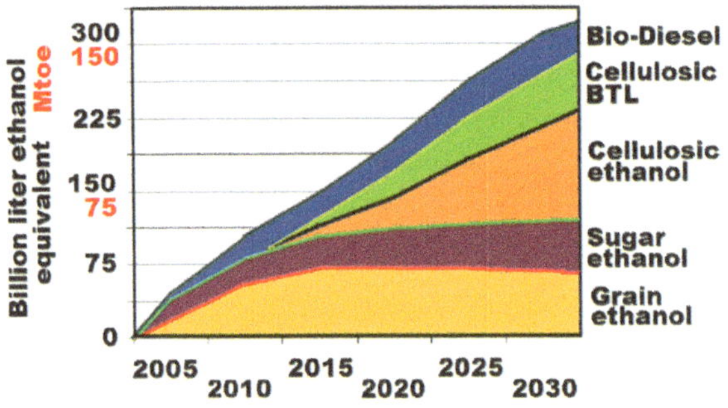

Fig. 3.2 World biofuel production by type (*Source* [3])

oil seeds produced biodiesel will be the major biofuel source over the next years, second-generation fuels from cellulosic materials and other non-food feedstocks are expected to acquire an increased market share by 2030 (Fig. 3.2) [3].

Figures 3.3 and 3.4 depict global production data in main producing regions. According to the OECD-FAO Agricultural Outlook (OECD-FAO 2012), global ethanol and biodiesel production increase is projected to slow down by 2021. Affected by policy mandates and renewable energy goals around the world, global ethanol and biodiesel productions are projected to reach, respectively, some 180 and 42 Bnl by 2021 [6].

Liquid biofuels—ethanol and biodiesel—provided 2.7% of the world's transport fuel in 2010, and consumed nearly 5% of world crop production in terms of energy content. Feedstocks for "first-generation" biofuels—which make up almost all the supply—are primarily major food crops, in particular maize, sugarcane, rapeseed, and soybeans (Fig. 3.5) [6].

The Biomass and Biofuel Energy Production in the European Union increases over time as shown in Fig. 3.6, and the share (%) of the total renewable energy increases too [7].

The biodiesel production in Greece over five years (2005–2010) increases from 12,000 to 17,000 tons as depicted in Fig. 3.7 [8].

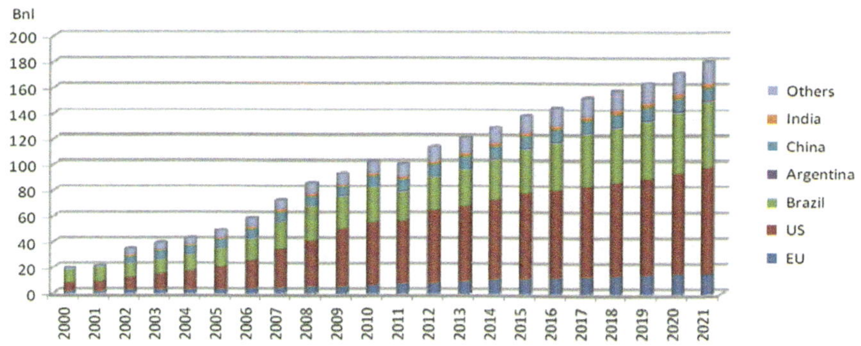

Fig. 3.3 Global ethanol production and projections to 2021 (*Source* [6])

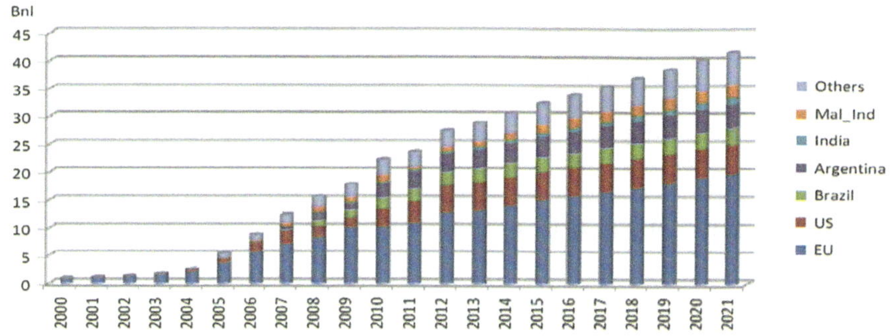

Fig. 3.4 Global biodiesel production and projections to 2021 (*Source* [6])

EU's reasons for promoting biofuel use are mainly environmental and geopolitical but also, secondarily, of a financial and social nature [9]:

- Environmental reasons aim toward the reduction of transport caused emissions, a contribution to achieving the Kyoto Protocol national targets and obligations and meeting the anticipated restrictions that will be imposed on pollutants emitted by car motors [9].
- From a geopolitical point of view, EU tries to secure fuel supply, to reduce imports, and to minimize Europe's and European countries' dependency on OPEC countries [9].
- On a financial level, promoting biofuel use creates new business and commerce opportunities around a high grossing sector such as the fuel sector, even in areas and countries so far not involved in oil production [9].
- Finally, social reasons include the potential to develop a new agricultural policy and to ensure new agricultural business areas, on local and national levels, thus creating jobs and utilizing rural lands in a more efficient way [9].

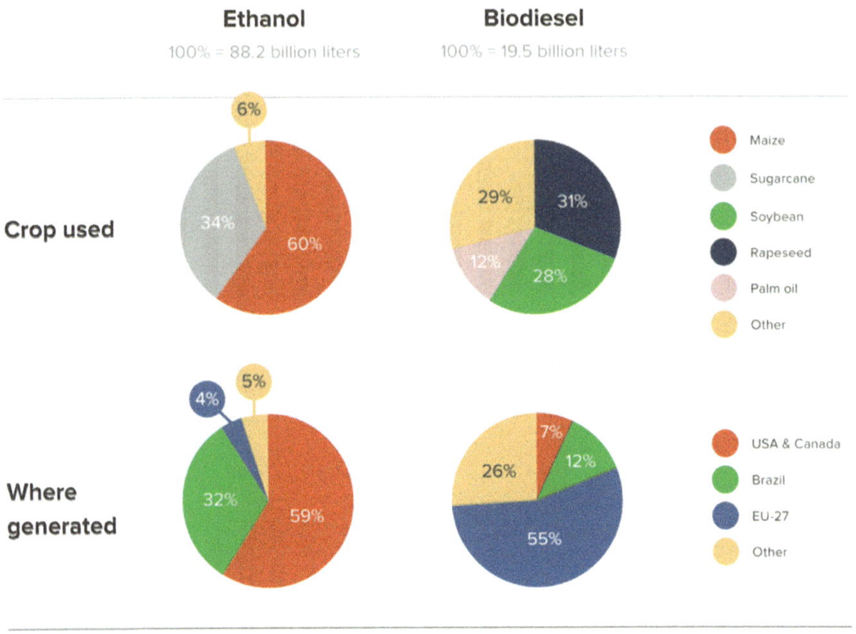

Fig. 3.5 Crop used for ethanol and biodiesel production and countries where they are generated (*Source* [14])

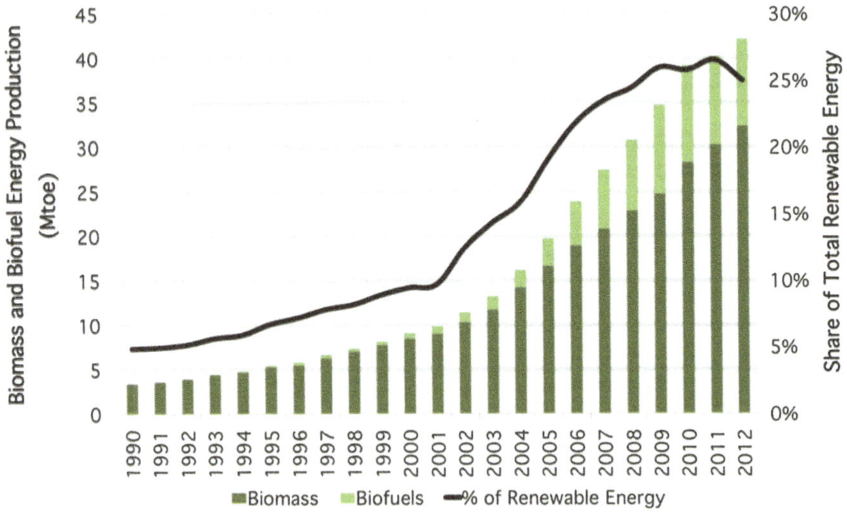

Fig. 3.6 Biomass and biofuels in the European Union (*Source* [7])

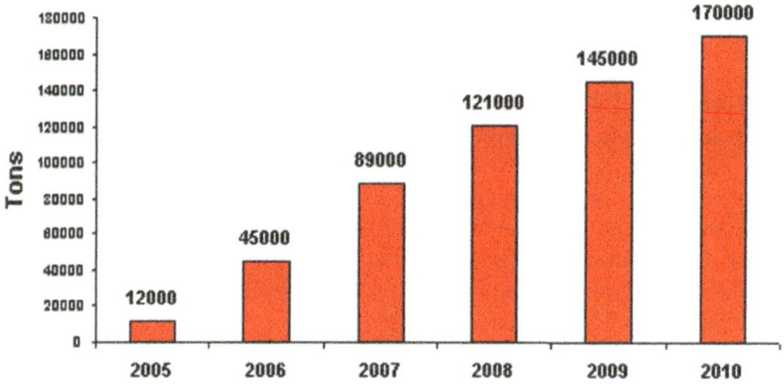

Fig. 3.7 Biodiesel production—provisions, Greece 2005–2010 (tons) (*Source* [8])

By 2020, EU aims to replace 10% of transport fuels with renewable energy fuels [10] and has imposed a 6% greenhouse gas intensity reduction on fuel providers [11]. However, in its communication of January 22, 2014, entitled "A policy framework for climate and energy in the period from 2020 to 2030" [12], the Commission proposed to abandon these two main biofuel targets after 2020 [13]. This follows concerns about land-use change emissions associated with biofuels which even brought about legislation amendments setting up an incentive scheme for biofuels that do not create additional demand for land [13].

The widespread use of first-generation biofuels raises serious questions about the opportunity cost of using all that land, water, and energy to produce fuel. That opportunity cost varies greatly depending on the feedstock used and the policy design. For instance, sugarcane-based biofuels provide far more energy in relation to inputs required than maize-based biofuels. An additional challenge to first-generation biofuels is the impact on food prices. There is widespread agreement that biofuels played at least a substantial role, among other factors, in the large increase in global crop prices since the mid-2000s. Inflexible mandates increase the price inelasticity of global demand for feedstocks in the face of supply shocks. Flexible mandates, as Brazil has used for sugarcane-based ethanol, are preferable [14].

To sum up, while first-generation biofuels are widely commercially produced all over the globe, they present several drawbacks, the primary one being that they are important food crops and their use for fuel can negatively affect food supply [1]. Furthermore, their cultivation is not free of environmental implications since they have intensive needs of one or more inputs such as land, water, fertilizers, pesticides [15]. In the future, cellulosic sources are expected to displace such crops as the major second-generation source of ethanol [1].

References

1. Janda K, Kristoufek L, Zilberman D (2012) Biofuels: policies and impacts. Agric Econ—Czech 58:372–386. http://www.agriculturejournals.cz/publicFiles/pdf. Accessed 12 Nov 2016
2. Sims REH, Mabee W, Saddler JN, Taylor M (2010) An overview of second generation biofuel technologies. Bioresour Technol 101:1570–1580. http://www.sciencedirect.com/science/article/pii/S0960852409015508. Accessed 12 Dec 2016
3. Spiess WEL (2011) Does biofuel production threaten food security. IUFoST Food Security Task Force.http://www.iufost.org/iufostftp/Does%20Biofuel%Production%20Threaten%20Food%20Security.pdf. Accessed 12 Nov 2016
4. Fulton L, Howes T, Hardy J (2004) Biofuels for transport: an international perspective. International Energy Agency, Paris, France. http://www.iea.org/textbase/nppdf/free/2004/biofuels2004.pdf. Accessed 21 Jan 2017
5. Koh LP, Ghazoul J (2008) Biofuels, biodiversity, and people: Understanding the conflicts and finding opportunities. Biol Conserv 141:2450–2460, Elsevier Ltd. http://www.sciencedirect.com/science/article/pii/S0006320708002954. Accessed 21 Jan 2017
6. Diop D, Blanco M, Flammini A, Schlaifer M, Kropiwnicka M A, Markhof MM (2013) Assessing the impact of biofuels production on developing countries from the point of view of policy coherence for development. Contract No 2012/299193, FWC COM 2011—Lot 1—Studies and Technical Assistance in all, Sectors, Final report, February 2013. https://ec.europa.eu/europeaid/sites/devco/files/study-impact-assesment-biofuels-production-on-development-pcd-201302_en_2.pdf. Accessed 12 Nov 2016
7. Boisvert W (2014) Harmonic destruction-how greens justify bioenergy's assault on nature. The Breakthrough Institute. http://thebreakthrough.org/index.php/journal/past-issues/issue-4/harmonic-destruction. Accessed 12 Feb 2017
8. Eco press (2012) Application of biofuels in Greece. http://ecopress-project.blogspot.gr/2012/03/blog-post_06.html. Accessed 14 Feb 2017
9. Ageridis G, Christou M (2006) Biofuels and environment throughout the life cycle. Biofuels and their contribution in the development of industry and agriculture, TCG/SCM, Workshop, Thessaloniki. http://www.cres.gr/kape/publications/pdf/eisigiseis/ageridis_biofuels.pdf. Accessed 12 Feb 2017
10. Renewable Energy Directive (2009/28/EC) http://eur-lex.europa.eu/legal-content/EN/ALL/?uri=celex%3A32009L0028. Accessed 19 Feb 2017
11. Fuel Quality Directive (2009/30/EC) http://eur-lex.europa.eu/legal-content/EN/TXT/?uri=celex%3A32009L0030. Accessed 19 Feb 2017
12. COM/2014/0015, Communication from the Commission to the European Parliament, The Council, The European Economic and Social Committee and the Committee of the Regions. A policy framework for climate and energy in the period from 2020 to 2030. http://www.ipex.eu/IPEXL-WEB/dossier/document/COM20140015.do. Accessed 19 Feb 2017
13. European Parliament News (2013) Renewable energy. http://www.europarl.europa.eu/RegData/etudes/fiches_techniques/2013/050704/04A_FT(2013)050704_EN.pdf. Accessed 12 Nov 2016
14. The new climate economy Report (2014) Land use. In: Better growth, better climate. http://newclimateeconomy.report/2014/land-use. Accessed 28 Jan 2017
15. Ziolkowska J, Simon L (2011) Environmental implications of biofuels—theoretical and empirical analysis for the EU and US. J Japan Inst Energy 90:177–181. https://www.jstage.jst.go.jp/article/jie/90/2/90_2_177/_pdf. Accessed 12 Nov 2016

Chapter 4
Biomass–Biofuels and Sustainable Approach

4.1 Biomass–Biofuels and Sustainability

Although biomass-produced energy is a major step toward the limitation of fossil fuel use and the subsequent environmental impact, it does not automatically imply that it falls under the category of sustainable approaches. Since it is already used as food and industrial raw material, using biomass as an energy source should meet the criteria of sustainability. In order to evaluate the biomass projects from a sustainability point of view, apart from environmental aspects, economic, social, cultural, and health aspects have to be examined as well. A major environmental condition to be met is the carbon balance over life cycle. The biomass chain could be characterized by carbon negative balance (net removal of CO_2eq. from the atmosphere) as well as carbon positive balances (net addition of CO_2eq.). These are determined by field practices, transport and processing technologies [1].

However, GHG emissions are not the only concern. A valid sustainability analysis applied to the bioenergy sector would have to account for social, environmental, and economic implications as shown in Figs. 4.1 and 4.2 [1].

Consequently, an approach could be equitable, bearable, or viable but only if it includes all three can it be described as sustainable. Biomass projects will not be completely successful unless they can ensure sustainable biomass supply, viable business conditions, and social support, as summarized (Table 4.1) [1].

RED 2009/28/EC directives have significantly altered the way biomass projects are evaluated. Initially, biomass territories were planned according to potential biomass values and later according to available biomass values. However, since not all available biomass can be sustainable, RED directives point out the necessity of "Sustainable Biomass" [1]. As mentioned before, biofuel benefits range far beyond the narrow energy security scope. They have a positive effect on the environment, boost economy and reduce imports and have a major social impact on the rural sector. Therefore, biofuel projects can fall in line with the sustainability concept of interlinkage and balance between economic, social, and environmental concerns [2].

A. Paschalidou et al., *Using Energy Crops for Biofuels or Food: the Choice*,
Green Energy and Technology, https://doi.org/10.1007/978-3-319-63943-7_4

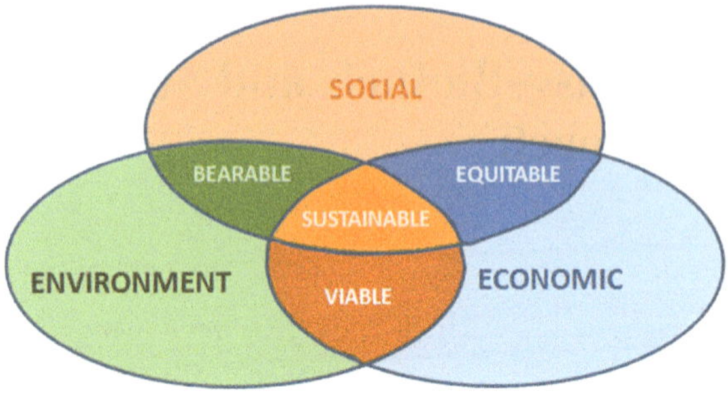

Fig. 4.1 General concept of sustainable approach (*Source* [1])

With the last Directive of sustainability criteria for biomass uses, the European Commission has introduced the most comprehensive and advanced sustainability scheme and the Member States are responsible to confirm and implement them for biofuels/bioliquids produced in the EU [1].

According to this scheme, a major factor in meeting the sustainability criteria is the land typology. Biofuels are not to be produced in high biodiversity value lands, which are geographically identified by the Commission. Moreover, raw material should not be obtained from primary forest, from designated nature protection areas, wetland, from areas with 10–30% canopy cover or peat-land [1]. Also, the Directive examines the biofuel coming only partly from non-renewable sources, indicating which percentage of fuel is renewable. The contribution of each energy source has to be taken into account on the basis of its energy content [1].

Finally, according to the Directive, biofuels feedstock from carbon-rich or bio-diverse land and food crop-based biofuels should be limited to a 7% share in the

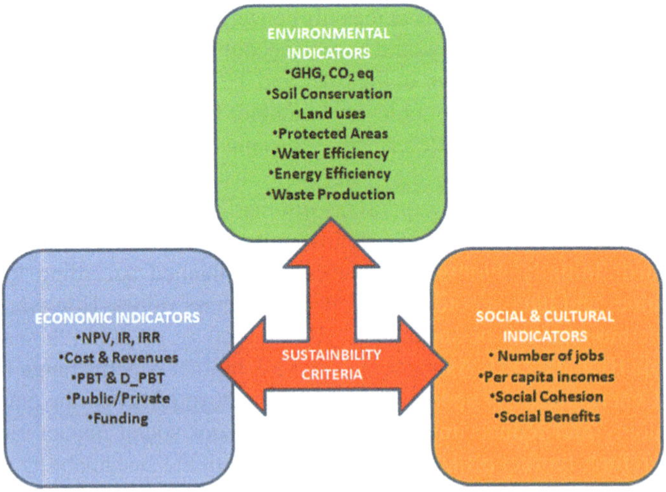

Fig. 4.2 General approach for a Bioenergy project (*Source* [1])

Table 4.1 Hierarchy of sustainability considerations of biomass projects (*Source* [1])

Sustainability criteria	Aspects evaluated
Ecologically sustainable and viable biomass supply	– Land availability
	– Water availability
	– Biodiversity
Commercially and technologically viable processing business	– Feedstock supply
	– Technology
	– Products and markets
License to operate	– Government directives
	– Community directives
	– Public consensus

transport sector. The Member States are obliged to guarantee compliance with these criteria (GHG, origin of feedstock) and European Commission (EC) has to report on compliance with environmental and social sustainability criteria of major biofuel exporting countries [3].

As stipulated in the Renewable Energy Directive, use of biofuels must result in an overall GHG saving of 35%, in order to qualify toward the 10% biofuels, target in the EU27 by 2020. This was set to rise to 50% from 2017 for existing production, and 60% for new installations from 2017. For plants already operating in January 2008, the new GHG requirement was set to start in April 2013. In an amendment to the Renewable Energy Directive in September 2015 (2015/1513), these deadlines were brought forward to 60% GHG saving for installation after October, 25, 2015, and 50% for existing installation after January, 1, 2018 [3].

It is worth noting that for these conditions to be enforced and observed, a strict certification system that ensures transparency needs to be established. Such a system should set objective quantified standards covering various sustainability criteria as well as direct and indirect impacts on the environment (water, biodiversity, etc.) and socioeconomic issues (food pricing, land availability, quality of life and social stability) [3].

4.2 Energy Planning—Sustainable Approach

Global energy is currently largely based on fossil fuels. The uneven distribution of fossil fuel reserves has resulted in a centralized energy model and the consequent dependency for the rest of the world. In addition, fossil fuel use has brought about significant economic development in the last century, to the benefit of oil producing countries mainly, but it has also created a vast economic sector. However, fossil fuel depletion, growing energy demands, environmental impact, and territorial conflicts are exposing the inherent weaknesses of the existing energy model, the biggest of which is a profound lack of sustainability. In order to take measures toward sustainable energy planning and energy diversification, policy makers have to incorporate security, efficiency, environmental, and social issues along with the undeniable need for a viable, realistic financial model that could generate local wealth. Therefore, since the energy sector represents an important financial area that provides people with an essential commodity, sustainable energy planning based on renewable energy sources entails the need to reconcile a multitude of criteria, often incompatible with each other [4].

Consequently, policy makers and energy planning in general have often relied on scientific methods that enable decision making and take into consideration multiple factors. Two decision making techniques, which are widely used are Delphi surveys and SWOT analysis. By using such techniques, it is possible to assess advantages and disadvantages of different energy sources or models, to highlight potential benefits and to anticipate obstacles. Most importantly, it enables future strategy planning as well as providing with alternative options [4].

Delphi techniques, a popular planning method, rely on a number of experts which evaluate a proposed course of action or a hypothesis. It is most often used as a part of a larger methodology. As a result, it has shown that the use of multicriteria decision analysis combined with expert judgement provides valid results, by supporting experts' opinion with objective data analysis. Furthermore, current software applications enable the use of an increased number of experts and allow planners to perform simulations and sensibility analysis in a quicker and easier way [4].

Finally, SWOT analysis is used as a comprehensive examination of an energy system or an alternative energy source. By presenting strengths, weaknesses, opportunities, and threats in a matrix, it is feasible to quantify and rank them, thus enabling decision making and planning. Moreover, this methodology facilitates the development of strategies that maximize on a system's strengths or remediate its weaknesses [4].

References

1. SEE, Energy Efficiency and Renewable Energy Sources—Support of Energy Policy at Local Level, Energy Supply, Manual Renewable Energy. http://ener-supply.eu/downloads/ENER_handbook_en.pdf. Accessed 15 Dec 2016
2. Demirbas A (2009) Progress and recent trends in biodiesel fuels. Energy Convers Manage 50(1):14–34. https://doi.org/10.1016/j.enconman.2008.09.001. http://citeseerx.ist.psu.edu/viewdoc/download?doi=10.1.1.463.9156&rep=rep1&type=pdf. Accessed 21 Jan 2017
3. Biofuels and Sustainability Issues, European Biofuels Technology Platform (EBTP). http://biofuelstp.eu/sustainability.html. Accessed 14 Feb 2017
4. Terrados J, Almonacid G, Aguilera J (2010) Energy planning: a sustainable approach, paths to sustainable energy, Artie Ng (ed). InTech. ISBN: 978-953-307-401-6. http://www.intechopen.com/books/paths-to-sustainable-energy/energy-planning-a-sustainable-approach. Accessed 12 Dec 2016

Chapter 5
Using Energy Crops for Biofuels or Food: The Choice

Since the development of bioenergy and the novelty of growing edible crops for non-food uses such as biomass, people have been preoccupied with a series of important questions: What are the short- and long-term consequences of such a dramatic change in the use of crops? What is the impact of this shift on food supply, food prices, and global nutrition? What is the cost that humanity might be asked to pay in order to minimize fossil fuel use?

These concerns have taken the form of long-standing controversy between two opposite viewpoints that essentially form the dilemma: "Should energy crops be used for fuel or for food?" [1]. This choice is not as simple as deciding, for example, whether to produce biodiesel or sunflower oil from sunflower or to produce bioethanol or sugar from sugar beet (Table 5.1):

Table 5.1 Basic energy crops for biofuel production and human food

Scientific name	Common name	Duration of the crop cultivation	Type of biofuel production	Food kind
Hordeum vulgare	Barley	Annual	Bioethanol	People and animal food
Secale cereale	Rye	Annual	Bioethanol	People and animal food
Triticum aestivum	Soft wheat	Annual	Bioethanol	People and animal food
Zea mays	Corn	Annual	Bioethanol	People and animal food
Sorghum bicolor	Sorghum	Annual	Bioethanol	People and animal food
Beta vulgaris	Sugar beet	Annual	Bioethanol	Sugar—food
Helianthus sp	Sunflower	Annual	Biodiesel	Sunflower oil
Glycine max	Soya	Annual	Biodiesel	People and animal food

A. Paschalidou et al., *Using Energy Crops for Biofuels or Food: the Choice*, Green Energy and Technology, https://doi.org/10.1007/978-3-319-63943-7_5

The biofuel debate is a broader issue that centers around the use of edible crops for biomass and the potential detrimental effect on the world. The two sides have come to be known as the antibiofuels lobby and the probiofuels lobby [2]. Their respective arguments range from valid scientific observations to media wars and even ethical issues [2].

On the one hand, the case against biofuels focuses on the following points:

A rapidly growing world population is already facing hunger problems. Nearly 60% of humans in the world are currently malnourished, so the need for grains and other basic foods is critical [3]. Using food crops to produce biofuels will further aggravate the problem in two ways: first by changing land-use patterns because land needed for food production would be diverted to energy crops, and second products would be destined for energy rather than food [4]. Both would result in food price increase and food insecurity, which will disproportionately affect the poorest people in developing countries.

Furthermore, the main energy crops—maize, soybean—are also feed crops, and their use for biomass would have a direct effect on the livestock sector. Second-generation biofuels use lignocellulosic biomass from wood or stalks which in developing countries serve in a major way as firewood and construction material [4]. Generally, land competition, including ILUC, will lead to deforestation, ecosystem destruction, and loss of biodiversity [2]. Consequently, using food crops for biomass raises major nutritional and ethical concerns [5].

While biofuels claim to be the "green" equivalent of fossil fuels, many of the social and environmental benefits of biofuels are not yet fully proven. In most cases, more fossil energy is required to produce a unit of biofuel compared with the energy that it produces [5]. Biofuels offer poor GHG benefits and a poor or negative energy balance [2]. Producing biofuels requires huge amounts of both fossil energy and food resources as well as water and other inputs [5]. Large-scale production of biofuels will increase soil erosion, decrease carbon stocks, put stress on water resources and on other ecosystem services [2].

Moreover, increased biofuel production may affect the quality of the remaining food plants in crop systems because it entails increased CO_2 emissions. This is reported to reduce the nutritional quality of major world foods, including wheat, rice, barley, potatoes, and soybeans since protein levels may be reduced as much as 15% [5].

On the other hand, biofuel defenders raise the following points:

Since large-scale biofuels production at global level is highly unlikely, there is sufficient land available to produce both food and a reasonable portion of biofuels without affecting food supply. This can be achieved with good and modern agricultural management practices and the utilization of marginal lands to the benefit of all parties concerned. Also, certain biofuel crops can be utilized to prevent soil degradation. Sustainable agricultural production systems, that produce both food and non-food products, already exist in many countries [2].

Although first-generation biofuels like biodiesel and bioethanol have been blamed for rising food prices, in fact, these prices are affected by a multitude of factors like weather conditions, oil price, growing demand from emerging economies, low investment in agricultural research, export/import policies, or even speculative investment [6]. Generally, it is argued that the use of different economic models and competing forms of statistical analysis makes it impossible to determine exactly how food prices are affected by biofuels and to separate their influence from other factors [1]. Besides, food deficit countries are highly unlikely to subsite food crops with fuel crops because biomass is not easily traded and local energy needs are still low. In more developed countries, governments can ensure the balanced land use and implement policies that protect food production [4].

In addition, bioenergy can spark off new investment in agriculture, bringing about economic development, creating jobs, and increasing productivity. Biofuels can play a significant role in modernizing and diversifying agriculture. Nearly 2000 million people live without access to modern energy systems. By providing more energy in developing countries, bioenergy can facilitate agriculture and can actually enhance food production. Generally, the social, economic, and environmental benefits of biofuels outweigh potential negative impacts, if good management practices are applied [2].

To sum up, it is apparent that the food versus fuel debate is a complex and multidimensional issue. In order to adequately determine the impact of biofuels on global nutrition, a broader and more in-depth analysis is required, to take into account all the different factors and facts that constitute the world energy systems [6].

References

1. HLPE (2013) Biofuels and food security. A report by the High-Level Panel of Experts on Food Security and Nutrition of the Committee on World Food Security, 13–14, Rome. http://www.fao.org/fileadmin/user_upload/hlpe/hlpe_documents/HLPE_Reports/HLPEReport-5_Biofuels_and_food_security.pdf). Accessed 22 Nov 2016
2. Rosillo-Calle F (2012) Food versus fuel: toward a new paradigm—the need for a holistic approach, Centre for Environmental Policy, Imperial College London, South Kensington Campus, London SW7 1NA, UK, International Scholarly Research Network, ISRN Renewable Energy, Vol 2012, Article ID 954180, 15 pages. https://doi.org/10.5402/2012/954180. https://www.hindawi.com/journals/isrn/2012/954180/. Accessed 12 Dec 2016
3. The World Health Report (2005) http://www.who.int/whr/2005/en/. Accessed 19 Feb 2017
4. Louise O (2006) Fresco Biomass for food or fuel: Is there a dilemma? Universiteit van Amsterdam Member of the Supervisory Board of Rabobank Nederland, The Duisenberg Lecture | Singapore. http://www.globalbioenergy.org/uploads/media/0609_Biomass_for_food_or_fuel.pdf. Accessed 12 Dec 2016
5. Pimentel D et al (2009) Food versus biofuels: environmental and economic costs. Human Ecol. https://doi.org/10.1007/s10745-009-9215-8. https://www.researchgate.net/publication/225484051_Food_Versus_Biofuels_Environmental_and_Economic_Costs. http://link.springer.com/article/10.1007/s10745-009-9215-8. Accessed 12 Dec 2016

6. Maunonen K (2009) Energy from field energy crops—a handbook for energy producers, Jyväskylä Innovation Oy & MTT Agrifood Research Finland. www.encrop.net, https://ec.europa.eu/energy/intelligent/projects/sites/iee-projects/files/projects/documents/encrop_international_handbook_for_energy_producers.pdf. Accessed 12 Dec 2016

Chapter 6
Methods (SWOT Analysis)

6.1 SWOT Analysis

A SWOT analysis is a strategic planning tool used to evaluate the strengths, weaknesses, opportunities, and threats involved in a project. It involves specifying the objective of the project and identifying the internal and external factors that are favorable and unfavorable to achieving that objective [1].

The aim of any SWOT analysis is to identify the key of internal and external factors that are important to achieving the objective. SWOT analysis groups key pieces of information into two main categories:

- Internal factors—The strengths and weaknesses internal to the organization or project.
- External factors—The opportunities and threats presented by the external environment [2].

The results are often presented in the form of a matrix which is shown in Fig. 6.1.

A SWOT analysis of any issue follows four basic steps: the external analysis to determine opportunities and threats, the internal analysis to identify the existing strengths and weaknesses, the generation of alternative strategies, and the formulation of a strategic choice [3].

6.1.1 External Analysis

The external analysis examines the existing threats or missed opportunities in the actual situation but also indicates possible trends and developments, ensuring, however, that the latter are rooted in realistic assessments, have a substantial effect,

A. Paschalidou et al., *Using Energy Crops for Biofuels or Food: the Choice*, Green Energy and Technology, https://doi.org/10.1007/978-3-319-63943-7_6

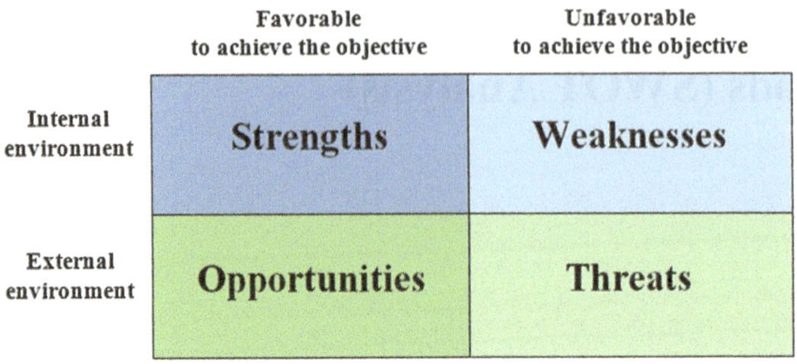

Fig. 6.1 General scheme of the SWOT tables (*Source* [1])

and are free of major speculations. An external condition or development that affects the project or issue under examination in a major way is defined as a threat if it has a negative influence and as an opportunity if the impact is positive [3].

6.1.2 Internal Analysis

The project or issue is examined for existing internal factors that critically determine its performance. Any internal asset that substantially facilitates the project's realization, exploits opportunities, or minimizes threats is identified as a strength. Weaknesses are internal characteristics that hinder the realization of the project and obstruct the exploitation of opportunities or make it vulnerable to threats [3].

6.1.3 Generating Alternative Strategies

Decisions based on SWOT analysis are formulated by combining strengths, weaknesses, opportunities, and threats in order to arrive at a number of alternative strategies, which may form the basis for further strategy formulation [3]. Figure 6.2 shows how they can be matched in order to arrive at four (or more) sets of alternative strategies. The alternative strategies are based on particular combinations of the four strategic factors:

SO strategies aim at using the strengths to exploit the opportunities.
ST strategies consider the strengths as a way to anticipate and avoid threats.
WO strategies attempt to take advantage of opportunities by overcoming weaknesses.
WT strategies are basically defensive and primarily act to minimize weaknesses and avoid threats [3].

Internal Factors / External Factors	**Strengths (S)** List of 5-10 internal strengths here	**Weaknesses (W)** List of 5-10 internal weaknesses here	
Opportunities (O) List of 5-10 external opportunities here	**SO Strategies** Generate strategies here that use strengths to take advantage of opportunities	**WO Strategies** Generate strategies here that take advantage of opportunities by overcoming weaknesses	**OT Strategies** Generate strategies that use opportunities to combat threats
Threats (T) List of 5-10 external threats here	**ST Strategies** Generate strategies here that use strengths to avoid threats	**WT Strategies** Generate strategies here that minimize weaknesses and avoid threats	
	SW Strategies Generate strategies that use strengths to mitigate weaknesses		**Synthesis**

Fig. 6.2 SWOT matrix: enhanced format (*Source* [3])

6.1.4 Formulation of Strategic Choice

Once alternative strategies are formulated, they are discussed and further analyzed. The general aim of this analysis is to choose ways to capitalize on strengths and take advantage of opportunities as well as overcome weaknesses and minimize threats. Gradually, strategic choices will emerge and the process of strategic choice formulation will result in one single strategy complete with planning for contingencies [3].

It should be noted that the factors displayed in the SWOT matrix (Fig. 6.1) refer to a particular point in time. However, external and internal factors may present minor or major changes over time. In order to account for the dynamic nature of environments, several matrices could be prepared at different points in time. For example, a SWOT analysis can start with a matrix representing the situation in the past, continue with an analysis of the present, and, perhaps most important, focus in the future [3].

6.2 The Use of SWOT Analysis

SWOT analysis is used to investigate the current state of energy crop cultivation for biofuel production or for food and identify the actions needed regarding economic viability, environmental impact, and social development. Therefore, the cultivation of crops for biofuel production or for food is considered as an action or project to be analyzed by SWOT in order to identify the opportunities and threats in the global external environment, and the strengths and weakness inherent to the project.

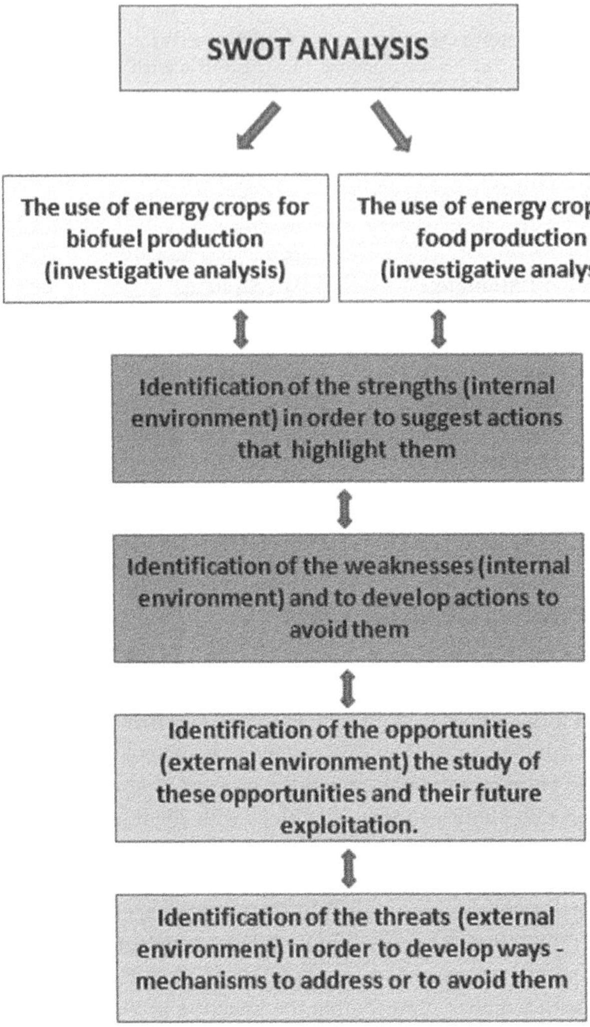

Fig. 6.3 Stages of the SWOT analysis of the two options

The analysis stages were the following (Fig. 6.3):

(a) The use of energy crops for biofuel production (investigative analysis):

- Identification of the **strengths** (internal environment) such as the energy supply and the economic impact of the biofuel production from energy crops in rural areas, as well its environmental impact. Identifying the strengths is necessary in order to suggest actions that highlight them and also develop ways to make the best use of these strengths.

- Identification of the **weaknesses** (internal environment) such as the possibility of affecting the quality of soil, air, water, and biodiversity and to develop actions to avoid them.
- Identification of the **opportunities** (external environment) such as the economic competitiveness with the fossil fuel and also the study of these opportunities and their future exploitation.
- Identification of the **threats** (external environment) such as the increase in food prices, in order to develop ways—mechanisms to address or to avoid them.

(b) The use of energy crops for food production (investigative analysis):

- Identification of the **strengths** (internal environment) such as food security. The identification of the **strengths** could be followed by proposal of actions to point them out and to develop measures for their exploitation.
- Identification of the **weaknesses** (internal environment) such as lack of improvement of economic indicators in rural areas and suggestion of countermeasures.
- Identification of the **opportunities** (external environment) such as the development of new and non-edible varieties only for biofuel production and their study and future use.

Table 6.1 Aspects of the energy crops

Environment	Economics	Social aspects
• GHG emissions	• Productivity and processing efficiency	• Benefits for smallholders
• Biodiversity	• Competitiveness and comparative advantage of the feedstock	• Income opportunities
• Soil conservation and soil quality	• Net energy balance	• Employment opportunities
• Water availability, use and efficiency	• State of commercialization/competitiveness with reference products	• Change in traditional use and knowledge
• Nutrient input requirements	• Market conditions	• Supply with modern energy as substitute for traditional bioenergy
• Pesticides requirements	• Price competition with other types of energy plants	• Energy security
• Soil remediation	• Policies	• Gender aspects
• Soil erosion control	• Crop yield	• Food security
	• Energy content	• Food and feed prices
		• Cultivation knowledge
		• Social and public acceptance

Source [4]

- Identification of the **threats** (external environment) such as the failure of reversing the climate change and to develop ways—mechanisms to address or to avoid these threats.

For the purposes of this analysis, a large variety of environmental, social, economic, and sustainability aspects were addressed (Table 6.1) [4].

References

1. Plieninger T, Bens O, Hüttl RF (2006) Perspectives of bioenergy for agriculture and rural areas. Berlin-Brandenburg Academy of Sciences and Humanities, Brandenburg University of Technology, Germany, 35(2), 125. http://www.researchgate.net/profile/Tobias_Plieninger/publication/233654571_Perspectives_of_bioenergy_for_agriculture_and_rural_areas/links/00b495210685d3ec55000000.pdf. Accessed 20 Nov 2016
2. Rode M (2005) Energetische Nutzung von Biomasse und der Naturschutz. Natur und Landschaft (80):403–412. http://www.nul-online.de/Archiv/Archiv/Energetische-Nutzung-von-Biomasse-Reststoffen,QUlEPTQxMDQ5MjcmTUlEPTgyMDMw.html. Accessed 22 Nov 2016
3. Groenendijk L (2002) SWOT analysis: strengths, weaknesses, opportunities, threats social science division/ITC/Enschede/ Netherlands. http://pgis-tk.cta.int/m05/docs/M05U04_handout_SWOT_Analysis_Groenendijk_2002.pdf. Accessed 22 Dec 2016
4. Rutz D, Janssen R (2012) Sweet sorghum as energy crop: a SWOT analysis. WIP Renewable Energies, Germany WIP Renewable Energies, Munich, Germany, 2012 SWEETFUEL reference number: WP6, Task 6.4, D6.5. http://www.globalbiopact.eu/images/stories/publications/swot_analysis.pdf. Accessed 26 Apr 2016

Chapter 7
Results and Discussion

7.1 Energy Crops for Biofuel Production—SWOT Analysis

7.1.1 Strengths

(a) **Development of rural areas**

One of the major advantages of biofuels is rural economic development. Biofuel production can boost agricultural economy and encourage agricultural diversification [1]. The subsequent rural economic growth will bring about additional capital inflows, create demand for goods and services that provide employment, and reduce rural–urban migration [2]. Moreover, the cultivation can involve small-scale farmers provided that the conversion facilities are located near the crop sources in rural areas [3]. In remote areas or marginal lands, where food production is non-competitive due to expensive inputs and lower prices, farmers can benefit from the production of biofuels, in rotation with food crops, thus covering the need for food, generated income, and animal feed [2]. Consequently, the use of land to produce energy crops can be considered as a possible solution to problems such as the land abandonment, rising unemployment, and development opportunities in rural areas [4].

The cultivation of energy crops will create a need for the development of new plant varieties. It will also bring about the improvement of cultivation methods and equipment that will support the production and storage of the new plants. This will boost the declining agricultural economy and will lead to the development of local farming industry [5].

A fact that should not be overlooked is that three billion people still rely on unhealthy and GHG-intensive traditional fuels like wood. In this case, energy crops can not only limit deforestation but also provide people with modern energy, such as ethanol for transport, light, and cooking and improve their life quality [6].

A. Paschalidou et al., *Using Energy Crops for Biofuels or Food: the Choice*,
Green Energy and Technology, https://doi.org/10.1007/978-3-319-63943-7_7

In Greece, where there are 6 million acres of abandoned farming lands, energy crops can have a significant financial effect. The recultivation of these lands with perennial plants of almost zero cost to produce biomass which will be used in thermal and electrical energy production will have significant economic and environmental benefits on local and national level [7]. The cultivation of the above-mentioned land will involve 120,000 agricultural holdings and 340,000 workers and employees in the primary sector. Based on the current producer's price (70 €/t biomass), it will have an added value of 400 million Euros as well as 1 billion Euros increase of GDP [7].

The added value throughout the whole production chain up to the retail sale of the final product is estimated at 2 billion Euros. Furthermore, the recultivation of these lands will lead to the creation of 1000–1500 new manufacturing businesses with 30,000 new job positions in the manufacturing and transport sector, wholesale and retail trade and in the sectors of production, trade and maintenance of biomass burners and boilers and so on. The estimated contribution to GDP will reach 3.2 billion Euros annually [7].

(b) **Creation of direct and indirect jobs**

Bioenergy projects involving energy crops can make a significant contribution to employment opportunities and the generation of direct and indirect jobs [4, 8]. Job creation in the bioenergy industry seems to be enhanced by the fact that it requires relatively little investment [9].

Job positions can be created at every stage in the biofuel production chain [10] (see Appendix Fig. A.1):

- Directly in the agricultural sector through the number of people required for the cultivation and/or collection of biomass [10].
- In the fuel supply chain, through people required for transport, storage, compression, etc., of the raw material [10].
- In the energy sector [10].

The number of job positions varies according to the applied technology. In domestic heating, jobs are related to the installation and periodic maintenance of the equipment. In larger building units, there is a growing need for operators and workers depending on the unit size [10].

According to EU's statistical data, biofuel production in the EU is one of the renewable energy technologies that can generate most jobs. In 2008, 42,000 people were employed in fuel ethanol production. Further, 225,000 jobs can be created if 10% of EU's petroleum consumption is replaced with ethanol [11].

Looking at all EU-28 Member States (Fig. 7.1), differences in the relation between turnover and employment between countries highlight potential developments in countries with low scores [12].

By comparing employment in manufacture and agriculture in the biofuels and bio-based chemical sectors (Figs. 7.2 and 7.3), it is obvious that employment in agriculture has a much higher share in biofuels than in bio-based chemicals [12].

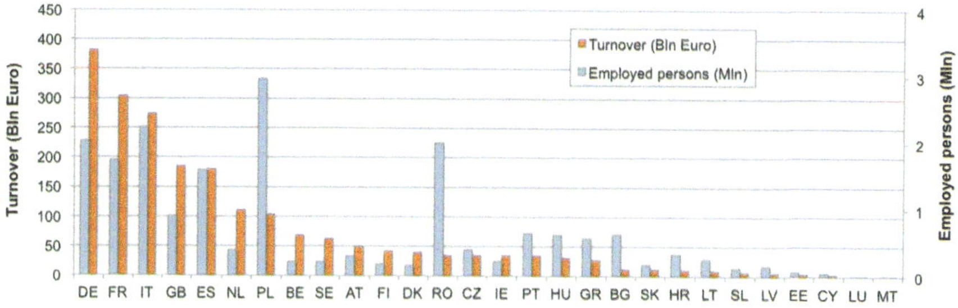

Fig. 7.1 Turnover and employment in the EU bioeconomy (2011) (*Source* [12])

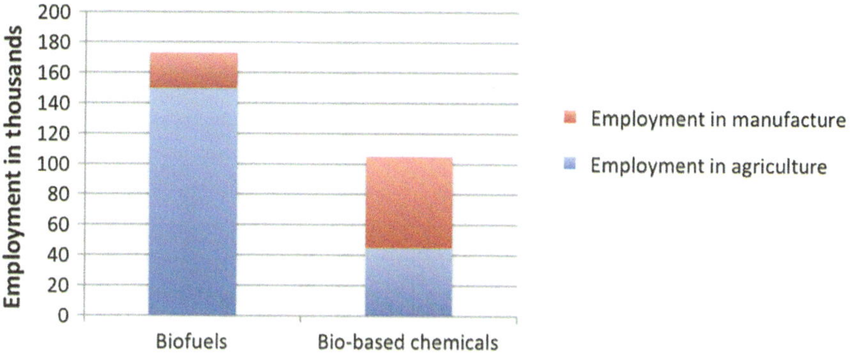

Fig. 7.2 Total employment (agricultural and manufacture) in biofuels and bio-based chemicals in EU-28, year (*Source* [12])

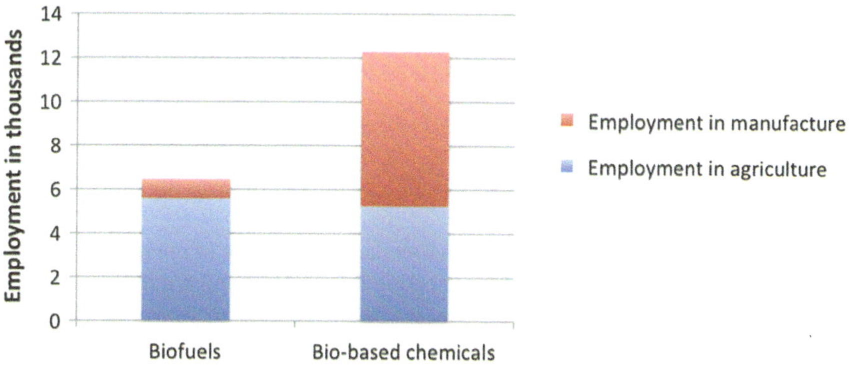

Fig. 7.3 Employment in biofuels and bio-based chemicals, per 1 million tones of feedstocks in EU (*Source* [12])

An overall examination of employment for biofuels and chemicals (Table 7.1) shows that at the manufacturing stage employment based on the same amount of biomass input is about 8 times higher for bio-based chemicals compared to biofuels, whereas employment in agriculture per ton of biomass is the same for both sectors.

Table 7.1 Overall comparison of employment in EU-28 (year 2011)

	Employment in agriculture	Employment in manufacture	Total employment (agricultural and manufacture)	Bio-based feedstock demand in million tones	Employment in manufacture in per 1 min t of bio-based feedstock	Total employment (agricultural and manufacture) in per 1 min t of bio-based feedstock
Biofuels	150,000	23,000	173,000	26.8	900	6500
Bio-based chemicals	45,000	60,000	105,000	8.6	7000	12,300

(*Source* [12]

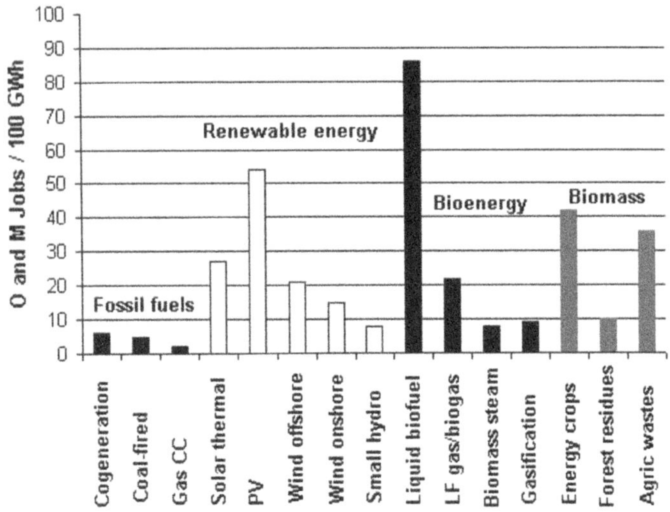

Fig. 7.4 Operation and maintenance jobs/MWh generated for a wide range of energy projects, both fossil fuel and renewable (*Source* [13])

Still, employment per 1 million tones of biomass is about twice as high for chemicals than for biofuels [12].

Employment for the operation and maintenance of a range of energy projects per MWh generated is given in Fig. 7.4 indicating the relatively high number of jobs in plant running or biomass supplying is associated with bioenergy projects [13].

In renewable energy sources like wind, hydro, and solar, employment consists mainly of manufacturing, installation, and maintenance, while on bioenergy projects, the biomass supply and delivering is an essential high labor component. Therefore, in bioprojects, there are more constant operation and maintenance jobs than construction jobs compared with other renewables (Fig. 7.5) [13].

Apart from the income generated by job creation, biofuel production offers opportunities for improving farmers' earnings. Many industrialized countries subsidy the export of surplus agricultural commodities. However, biofuels' production, which requires many of these crops as inputs, can be considered as a chance to domestically absorb agricultural surplus without the need for subsidies. This increased demand

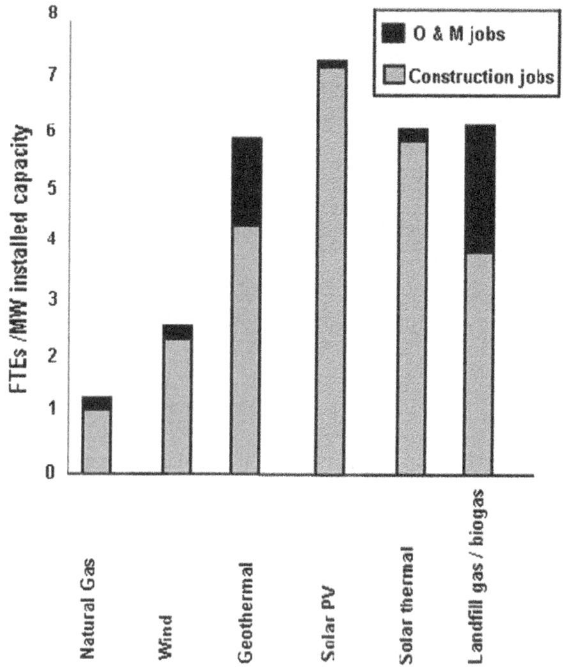

Fig. 7.5 Construction jobs and operation and maintenance jobs per MW of installed power-generating capacity for several renewable energy technologies compared with natural gas (*Source* [13])

for agricultural commodities can lead to price increase of agricultural commodities and therefore augment farmers' income [3].

(c) **Diversity of energy supply**

Renewable energy sources (wind, solar, hydro, energy from the oceans, geothermal, biomass, and biofuels) are alternatives to fossil fuels and enhance the diversification of energy supply [14, 15]. In contrast to fossil fuels, which have specific limited reserves, renewables can provide sustainable energy services based on the recurrent use of domestic resources, particularly if their costs decline and their performance characteristics continue to improve [16].

Of the leading energy alternatives to fossil fuels, biomass is most cost-effective and practical and therefore offers the most realistic and sustainable strategy. Biomass energy comes at low capital cost and offers a realistic and sustainable alternative to fossil fuels. It is versatile and can also be used to produce biomethane to replace natural gas, liquid biofuels to replace gasoline and diesel fuel, and as source material for biochemicals and bioplastics (Fig. 7.6) [17].

Figure 7.7 represents the amount of investments needed, according to BNEF, in order to avoid the most severe consequences of climate change, which is marked by an increase of more than 2 °C. Blue lines stand for required investments (in billions of dollars per year), and red lines depict current investments [18].

The positive effect of diversification in energy supply is the main reason behind EU's policy regarding renewable energy resources. The target to be reached by 2020

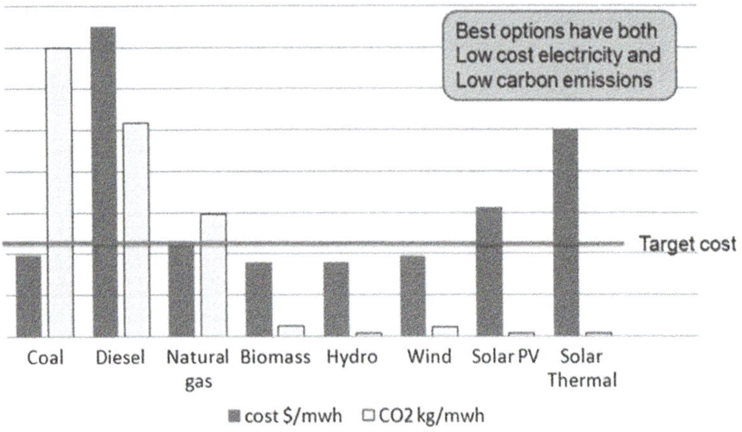

Fig. 7.6 Electricity cost and carbon dioxide emissions per kilowatt hour (*Source* [17])

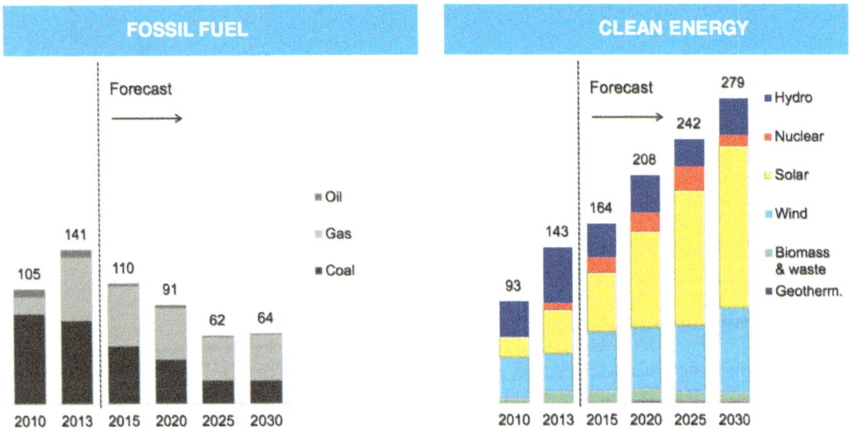

Fig. 7.7 Power generation capacity additions (GW) Bloomberg New Energy Finance (*Source* [18])

for the EU is a share of 20% energy from renewable sources in gross final consumption, and Member States have agreed on a new target of at least 27% by 2030. Significant steps toward meeting this target have already been made. As can be seen in Fig. 7.8 in 2014, the share of renewables was 16.0%, almost double that of 2004 (8.5%) [19] (see Appendix Table A.2 and Fig. A.2).

(d) **Independence from fossil fuel**

Most countries rely heavily on fossil fuel to cover their energy needs. However, fossil fuel supply is highly centralized, and prices are prone to fluctuation, affected by territorial conflicts. Consequently, many countries, which are dependent on energy imports, are in danger of supply disruptions and open to energy security risks. In 2000, OECD countries had to import more than half of their energy supply, and this

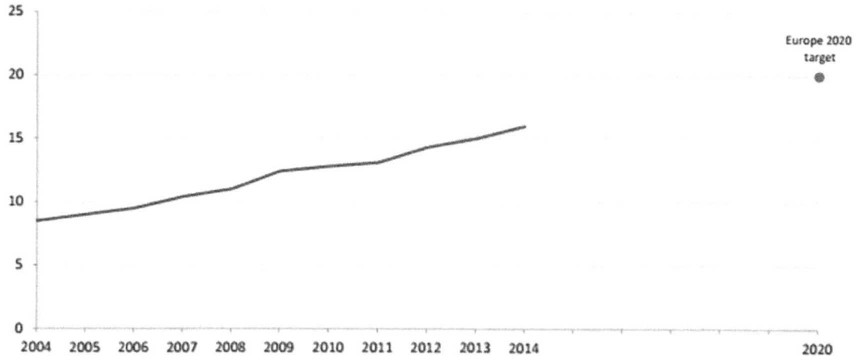

Fig. 7.8 Share of energy from renewable sources in the European Union (in 5% of gross final energy consumption) (*Source* [19])

is expected to rise by 2020 [3]. Renewable energy sources like biomass and biofuels can contribute to the diversification of energy supply, particularly in developing countries and rural areas, and reduce dependence on unreliable and volatile fossil fuel markets, particularly oil and natural gas [20, 15].

The introduction of biofuel can bring about the decrease of international fuel prices by between 1.07 and 1.10%, as well as the reduction of fossil fuel quantity consumed by oil-importing countries by between 0.3 and 0.7% [21].

(e) **Contribution to the reduction of the greenhouse effect**

Global warming has become widely accepted as a serious threat to the environment and to the planet. It is mainly caused by greenhouse gases which increase the percentage of CO_2 in the atmosphere. Fossil fuels add below ground carbon to the above ground cycle. According to NASA, the amount of carbon that fossil fuel is estimated to add to the atmosphere each year is 5.5 billion metric tons worldwide (see red graphic) (Fig. 7.9). This addition of subterranean carbon to the atmosphere is termed as "carbon positive" [22].

Most forms of renewable energies, solar, wind, hydroelectric, tidal, and geothermal, neither add nor reduce the amount of above carbon so they are "carbon neutral" (see gray graphic) (Fig. 7.9). However, energy crops used for biomass convert atmospheric carbon to glucose through photosynthesis. Consequently, bioenergy can be considered as the only carbon-negative energy source (see green cycle) (Fig. 7.9) [22].

Furthermore, even if the released carbon from biomass burning is taken into consideration, bioenergy can still be considered carbon neutral because CO_2 released into the atmosphere after combustion is absorbed by the plant during photosynthesis (Fig. 7.10) [5, 23, 24].

Figure 7.11 illustrates how biomass energy is carbon neutral and renewable [25].

However, production and combustion processes contribute to a net carbon emission for every biofuel despite the fact that recent improvements in biofuel production technologies have reduced carbon emissions. Fig. 7.12 shows the current biofuel car-

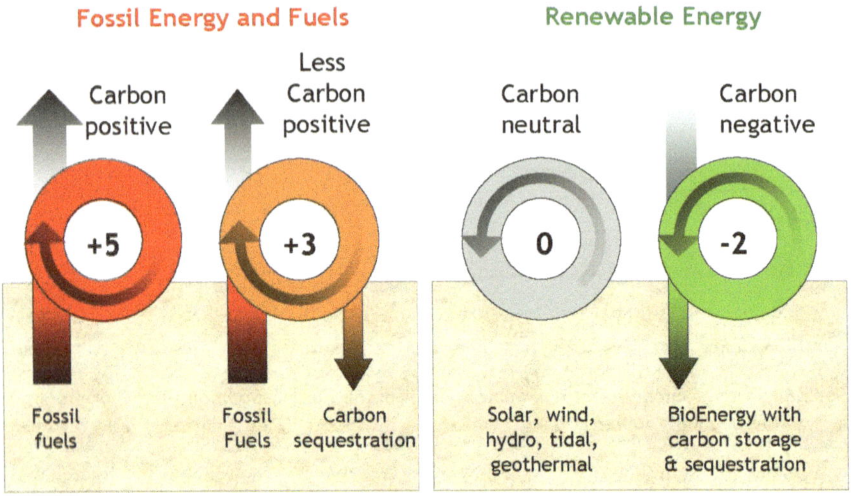

Fig. 7.9 Carbon footprint comparison (*Source* [22])

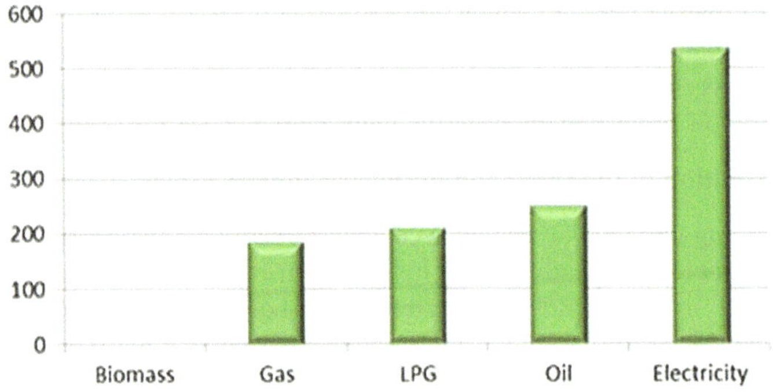

Fig. 7.10 CO_2 emissions (g per kWh for biomass, compared to other fuels (*Source* [5])

bon intensity statistics resulting from different feedstock and technology, compared to traditional fossil fuels [24].

Moreover, substituting gasoline or diesel with biofuels can reduce GHG emissions by 31% for bioethanol, 54% for biodiesel, and 71% for cellulosic ethanol according to studies carried out over the past 15 years (Fig. 7.13) [26].

Finally, according to the Intended Nationally Determined Contribution (INDC) Scenario, the world's prevalent fossil fuel consumer is the power sector, which is responsible for much of the energy-related CO_2 emissions. Substituting fossil fuels with renewables can help reduce these emissions (Fig. 7.14) [27].

(f) **Reduction of soil erosion during the replacement of energy fields by farmland**

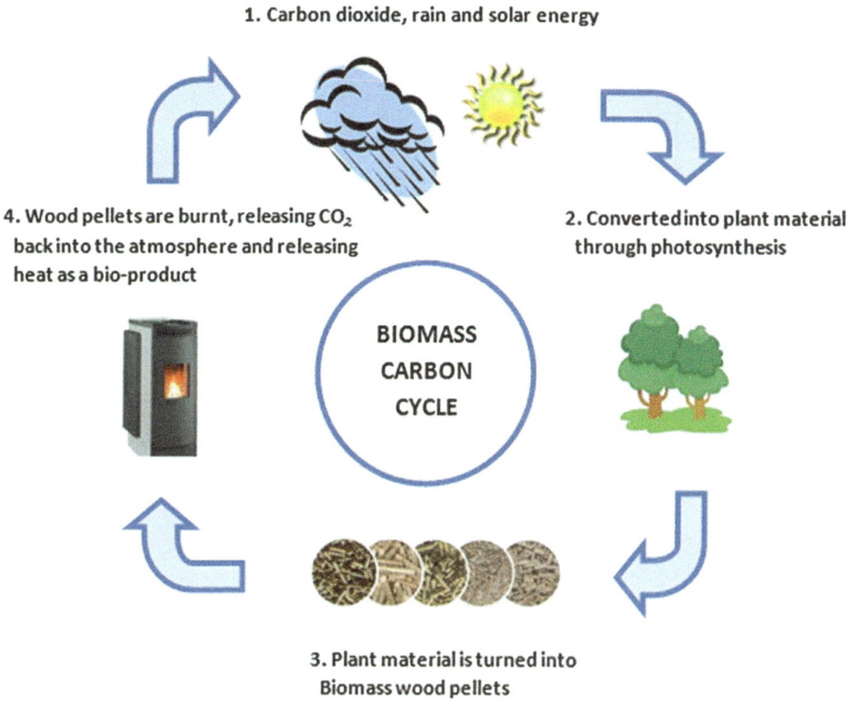

Fig. 7.11 Biomass carbon cycle (*Source* [25])

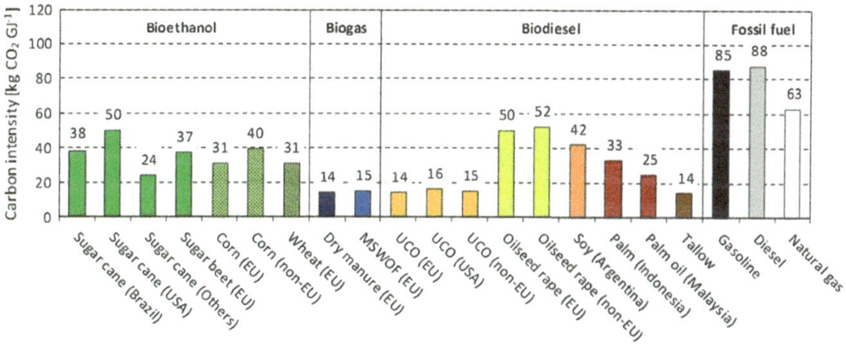

Fig. 7.12 Carbon intensity of biofuels resulting from different feedstock and technology, compared to traditional fossil fuels (*Source* [24])

Agriculture and farming, especially intensive farming, can have a detrimental effect on soil quality. There are energy crops, however, that can reduce soil erosion. Biofuel plants such as jatropha and pongamia, which grow on marginal lands, can improve soil quality and coverage and reduce erosion, while their oil cakes can provide organic nutrients for improving soil [2].

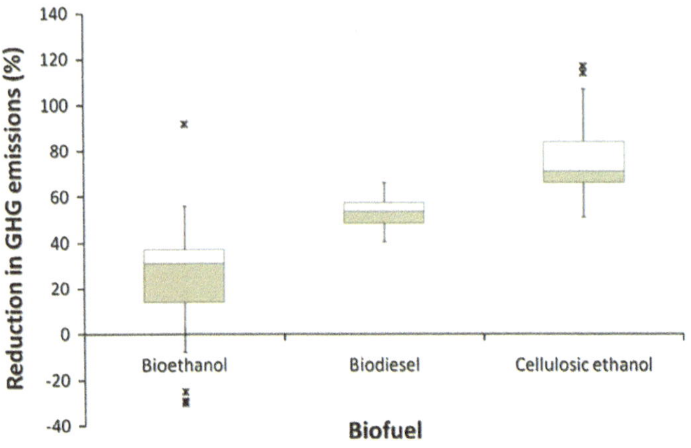

Fig. 7.13 Estimates of net reduction in greenhouse gas (GHG) emissions from the displacement of gasoline or diesel use in transport by different types of biofuels, including bioethanol (*Source* [26])

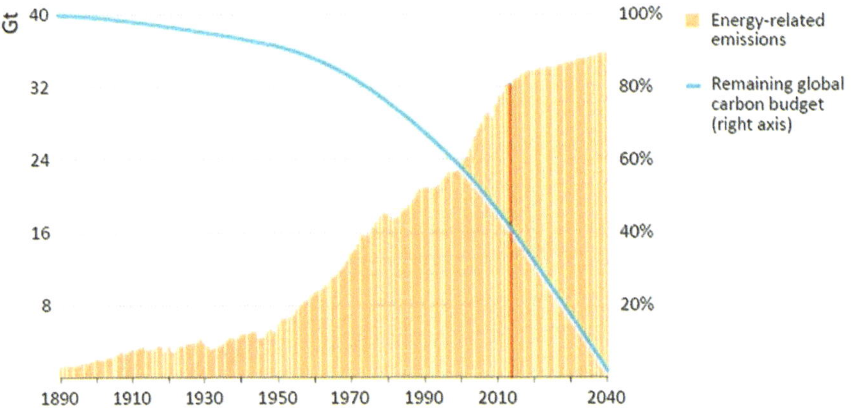

Fig. 7.14 Global energy-related CO_2 emissions in the INDC Scenario and remaining carbon budget for a > 50% chance of keeping to 2 °C (*Source* [27])

Moreover, extensive root systems of perennial grasses can potentially bind soil. This would prevent erosion and aid in filtering heavy metal contaminants from wastewaters [28]. Consequently, bioenergy improves soil structure and reduces erosion compared to intensively managed land [29, 30].

Several studies indicate that water-induced soil erosion will be reduced, particularly when cardoon, miscanthus, or switchgrass replace conventional annual crops [31]. Bical energy reports a high reduction in the rate of soil erosion when cereals are replaced by perennial grasses [32].

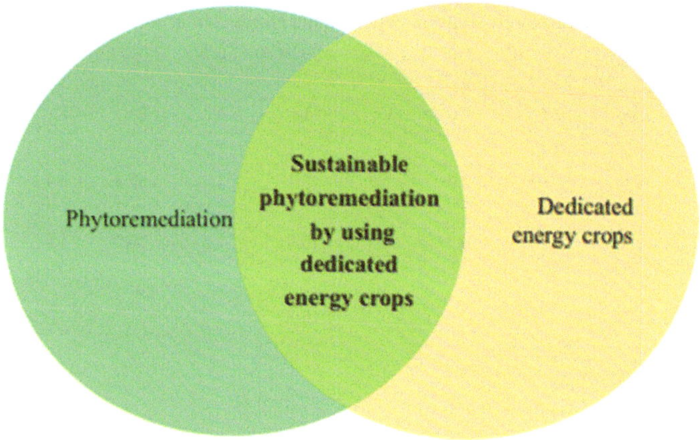

Fig. 7.15 Linking phytoremediation with dedicated energy crops to accomplish the "sustainable phytoremediation" (*Source* [33])

(g) Soil remediation

The increased awareness regarding environmental protection and human health has made both scientists and politicians realize the necessity to decontaminate polluted sites and especially soil, since this has a significant effect on the food chain. The process though which plants are used to remove harmful substances from the soil is termed phytoremediation and has gained much popularity over the last 20 years. A number of companies are engaged in phytoremediation of various contaminated sites, and several phytoremediation projects have been successfully completed all over the world [33]. Phytoremediation covers a wide range of pollutants like inorganic chemicals including heavy metals [34, 35], metalloids [36], many organic substances [37] including persistent organic pollutants [38] and radioactive elements [39, 40].

A wide range of energy crops is suitable for this purpose, thus achieving the dual effect of benefiting the environment while generating income from biofuels. Moreover, the use of metal-accumulating bioenergy crops, especially non-edible perennials to decontaminate heavy metal polluted soils, is a sustainable approach to phytoremediation [41, 35]. A good choice for the simultaneous bioenergy production and phytoremediation is dedicated energy crops or short-rotation crops (Fig. 7.15) [33].

Although linking phytoremediation with energy crops is a sustainable strategy, it can be achieved by involvement of practitioners, scientists, policy makers, phytoremediation companies, farmers, etc. [42, 43]. Several countries are constantly exploring the potential energy crops for biodiesel production based on their phytoremediation capability. In this direction, numerous studies regarding energy crop-based phytoremediation have been conducted worldwide (Table 7.2), but there is still ground for further research into potential energy crops for phytoremediation purposes in different types of contaminated lands [33].

Table 7.2 Results from phytoremediation experiments conducted at AUA

Normal values in plant tissues	0.02–7	0.01–0.2	0.02–1	5–20	20–1000	0.1–5	0.2–20	0.0001–0.2	1–400
Plant species	Ac	Cd	Co	Cu	Mn	Ni	Pb	Sb	Zn
Giant reed		**9.0**				**12.5**			
Kenaf	2.6	**10.0**	0.2	5.6	47.1	1.0	**65.8**	**3.4**	281.2
Cardoon		**169.3**				**342.3**			
Jatropha	1.3	0.2	0.2	2.4	107.1	1.3	**26.7**	**1.6**	86.1

Source [119]

7.1.2 Weaknesses

(a) **Effects on quality of soil, water, and biodiversity**

While agriculture's primary aim is to cover food supply and generate income for farmers, it must not be overlooked that agricultural landscapes serve another but equally important purpose, which is their immediate effect on the environment. They regulate water quality, sequester CO_2 and host beneficial insects, and other wildlife. Therefore, sustainable agriculture needs to balance those two roles and provide goods without forfeiting its environmental and social benefits and definitely without adding to the pollution [44].

The cultivation of energy crops for biomass can potentially have detrimental effects on biodiversity and landscape conservation [45]. Although traditional agriculture may be carried out on a small scale or even individually, bioenergy is often carried out on a large scale with definite financial goals [9]. This large-scale production could lead to specific problems, usually associated with intensive farming: For example, crops like corn or soybeans might be responsible for more pollution than the oil they are supposed to replace, due to large amounts of fertilizer, land, and water that are used for their cultivation [46, 47].

Generally, replacing crops with biofuels can result in industrial monoculture and biodiversity loss. Increased water requirements, for irrigation as well as processing crops into biofuels, reduce water flows. This, together with the increased use of agrichemicals, can discharge organically contaminated effluent, which could lead to the deterioration of water quality and affect downstream ecosystems. Moreover, it can reduce soil fertility and generally degrade land [2, 3, 13].

(b) **Dependence on land availability**

Biomass production has gained acceptance and popularity over the past years. This increased demand for biomass has resulted in increased demand for land on which to cultivate it. However, this pressure can lead to direct and indirect changes in land use that present a number of drawbacks both for developing as well as for developed countries [48]. Direct land-use change occurs when biofuel crops are grown in non-agricultural lands or diverse agroforestry systems. Moreover, if land previously used for food or feed is directed to biofuels, this would mean food crops

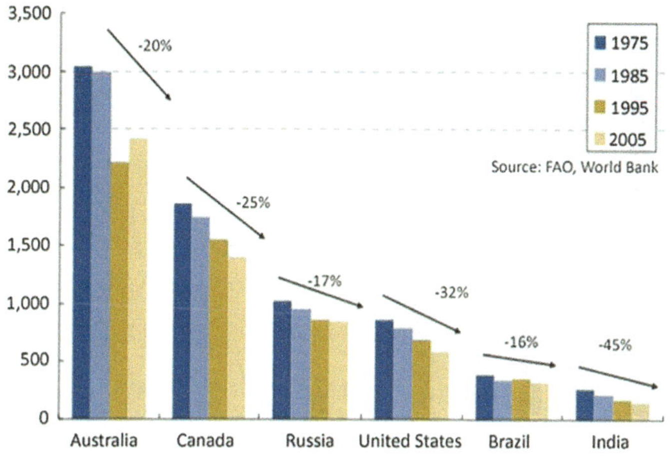

Fig. 7.16 Arable land per capita is decreasing (hectares per thousand people) (*Source* [51])

Table 7.3 Available arable land for bioenergy crop production in EU22 in 2010, 2020, and 2030

Year	Available arable land for bioenergy crop production in EU-22 (EEA 2006) (Mha)
2010	12.97
2020	16.17
2030	19.27

Source [48]

will have to be displaced and at least partially restricted to previously non-cropland such as grasslands and forests. This displacement is termed indirect land-use change (ILUC) [49, 26, 50].

EU relies on biofuels to meet its greenhouse gas reduction targets. In order to reach the 2020 GHG reduction target, EU plans to import 50% of the required biomass, since it is estimated that there is not enough cropland in Europe [24].

As can be seen in Fig. 7.16, the phenomenon of diminishing arable land is global. Less arable land per capita results in intensification of agricultural practices in order to achieve the necessary yields, with the subsequent environmental effects [51].

Table 7.3 summarizes the results of a study conducted by the EEA in 2006, regarding the amount of arable land in Europe that could be dedicated to energy crops by 2030 [48] (see Appendix Table A.1). According to this study, it could reach 19.3 Mha which represents 12% of the total utilized agricultural area (UAA) in 2030. In addition, up to 5.9 Mha could be released from grasslands and olive grove categories. According to estimates, Europeans already occupy more land for agricultural purposes than is available globally per capita [48].

Using information from the Energy Information Administration Reference scenario for the year 2030, Fig. 7.17 illustrates the total land-use intensity associated with various electricity production methods, as predicted for 2030 [52].

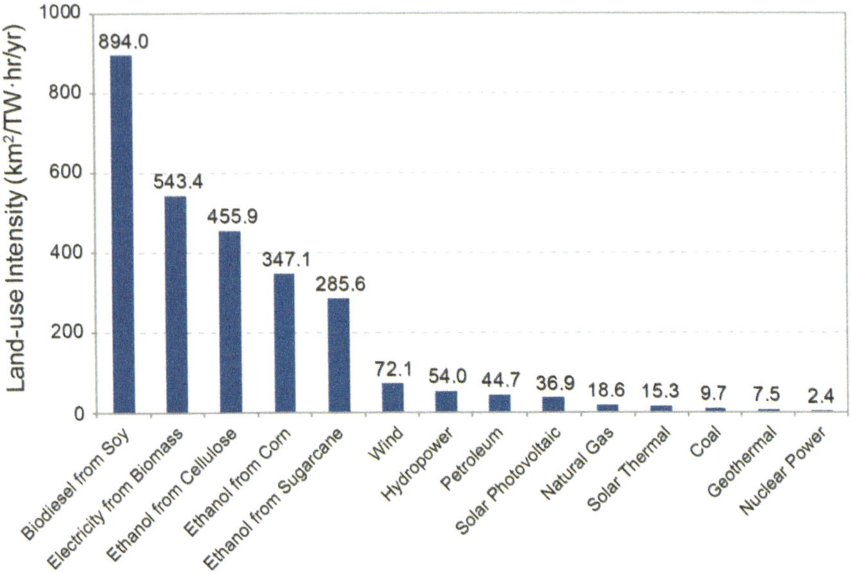

Fig. 7.17 Projected land-use intensity in 2030 (*Source* [52])

Although developing countries have large areas of land better suited for biofuel production than for food crops, the need for large amounts of cropland to produce energy crops is increasing land-grabbing, with negative socioeconomic impacts on local communities. Moreover, high demand and rapid expansion of biofuel production could increase conflict over land rights. Industries may move to appropriate land, especially if land tenure systems are weak [2, 24].

Moreover, this shift in land-use patterns has serious consequences on food and feed, with dangers varying from increased prices and limited access to nutrition to actual food shortage. These drawbacks will disproportionately affect poorer regions [53, 23].

Furthermore, the cultivation of marginal lands entails unpredicted damage to livestock grazing areas or wildlife corridors and the ensuing endangerment of sensitive ecosystems [54]. In tropical and subtropical developing countries, biofuels lead to deforestation which is regarded as the greatest threat to terrestrial biodiversity as forests are the most biologically diverse terrestrial ecosystems [55].

Another significant effect of land-use change is the increase of net GHG emissions. It is estimated that deforestation and forest degradation were responsible for 11% of total global GHG emissions [56]. Land-use change may also in turn contribute to GHG emissions from the loss of carbon stored in above- and belowground biomass when land is cleared and from the loss of the carbon sequestration service of converted land uses [26].

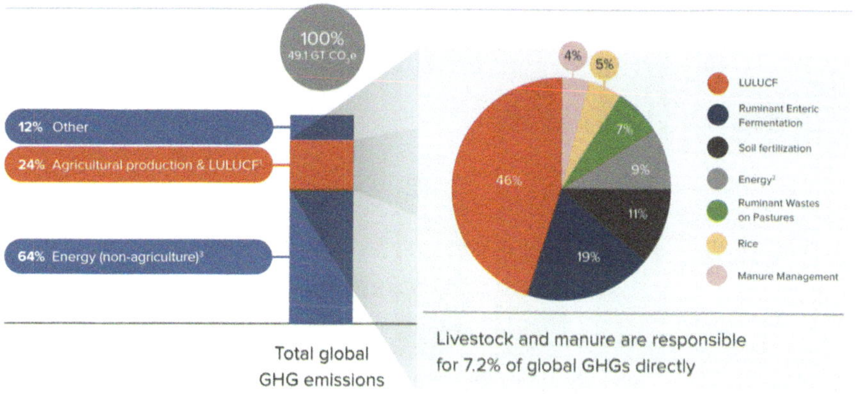

Fig. 7.18 Global AFOLU greenhouse gas emissions by sub-sector (2010) (*Source* [56])

Agriculture, forestry, and other land uses are themselves major producers of GHG emissions (Fig. 7.18). Biofuel production may multiply and intensify emissions induced by agriculture due to the land-use change [56].

(c) **Effects of land-use change**

Biofuels have been proposed as a sustainable alternative to fossil fuels, capitalizing on the fact that fossil fuel reserves are not limitless. However, biofuel production is highly dependent on available land, which, in turn, is not limitless either. On the contrary, various land uses which compete with each other present a serious territorial limit. Humanity relies on arable land for food and feed and for biofuels at the same time, but also depends upon non-arable land, forests, and natural ecosystems for biodiversity and environment conservation. These contradictory purposes may prove fatal for biofuel production, since global food production already utilizes about 5 Gha of the 10 Gha usable land surface available [57, 58].

The European Commission (2008) calculated that 17.5 million hectares of land would be required to reach the 10% biofuel target, which would amount to about 10% of the total Utilized Agricultural Area (UAA) in EU27. The OECD (2006) is less optimistic and estimates that about 45 million hectares of land are required to reach the EC-targets by 2020 [59]. This may cause adverse effects on biodiversity as it may lead to the further intensification of existing land uses, in both agricultural and forest lands, but also the conversion of non-cropped biodiversity-rich land into cropped or forest area [59].

(d) **Disease and Pest sensitivity during the production of primary source**

Global agriculture faces major challenges from pests, pathogens, and weeds resulting in yield losses in the preharvest period but also in transport, preprocessing, storage, processing, packaging, marketing, and plate waste [55].

Although current research on energy crops is focused almost exclusively on selection of high performing, varieties, and development of optimum agronomic practices,

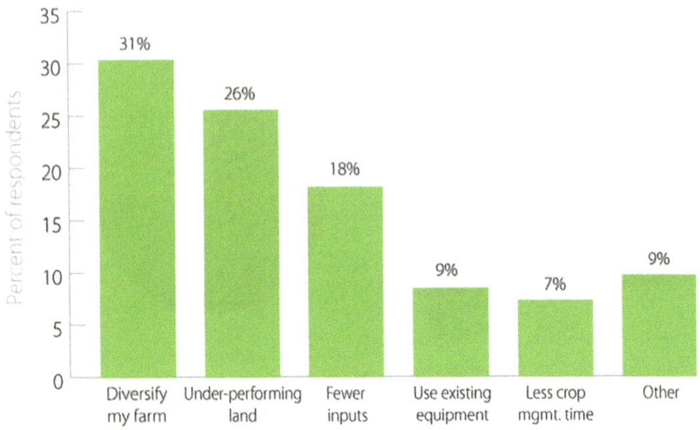

Fig. 7.19 Most appealing aspects of energy grasses (*Source* [63])

diseases may pose a threat to successful establishment and economics of bioenergy crop production. Especially since biofuels tend to be cultivated as large crop monocultures, the ensuing limited genetic diversity reduces local biodiversity levels and natural antagonists of pest and pathogen species and substantially increases the potential severity of disease outbreaks. Moreover, new disease threats are likely to emerge alongside existing diseases as energy crop monocultures become commonplace worldwide [60].

(e) **Farmers' lack of knowledge or potential reluctance to accept dedicated energy crops**

Motivating farmers to change production systems and choose biofuel cultivation is a complex issue that involves a lot of factors such as potential income, knowledge, legal conditions as well as traditional values. According to the relevant literature, there are six types of barriers, namely agreement, knowledge, technological, economic, social, and political, and administrative barriers, which are hindering increased energy crop cultivation [61].

Farmers' perspectives of bioenergy were researched in Ireland, and the results show that the degree of interest in adopting energy crop cultivation was not directly correlated with the degree of their knowledge about the crops. In other words, lack of knowledge on energy crop production is not the primary reason for a lack of interest. This was further supported by Research from Tennessee, USA, about switchgrass [62].

According to US survey, growers said that the ability to diversify their current operations was the most appealing benefit of energy grasses (Fig. 7.19). They were also interested in making better use of their marginal land and spending less time, money, and resources on crop management [63].

In general, strong financial incentives are probably the main reason for farmers to change their traditional land uses or production methods in order to grow

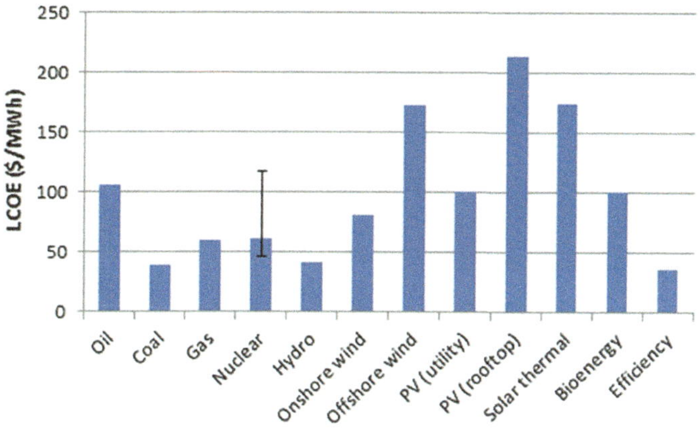

Fig. 7.20 Average global internalized costs of new builds of different electricity options (*Source* [64])

energy crops or to utilize their wastes for energy purposes. In contrast, education and awareness-raising about energy crops are likely to have a much smaller effect. In developing countries, cultural factors also have a strong influence on the general unwillingness to break farming traditions. Changing attitudes are unlikely to occur unless farmers can perceive personal cobenefits such as increased profitability, time saving, cost reductions, improved animal health, increased soil fertility, and, perhaps most important, being able to undertake less arduous tasks [13].

(f) **Production cost**

One of the main reasons why biofuel production has not reached the wide scale of fossil fuel use is their higher economic costs compared to conventional fuels, which is reported to be even double in some cases. The accumulated costs of planting, irrigating, input adding, harvesting, transporting, and refining are rendering biofuels more expensive [3]. Moreover, owing to the long term of use of fossil fuels, there is an already established infrastructure, while biofuels require investments in new plants. The average global internalized costs of new builds of different electricity options are presented in Fig. 7.20. It is apparent that conventional technologies (coal, gas, nuclear, and hydro) remain the most economical options on a global basis [64].

Also, the average global internalized costs of different transportation energy options are shown in Fig. 7.21. Oil is still by far the most economical fuel for transportation purposes [64].

It is clear therefore that biofuels are generally not as cost-effective as fossil fuels. However, biofuel production costs tend to differ depending on the type of biofuel, the country of provenance, and the technology used. There are also differences depending on the feedstock used. Corn, for example, is more expensive and produces less bioethanol per hectare than tropical crops such as sugarcane [3].

Figure 7.22 presents a comparison of current net production costs of liquid biofuels in various regions of the world. Although the production cost of bioethanol in the EU

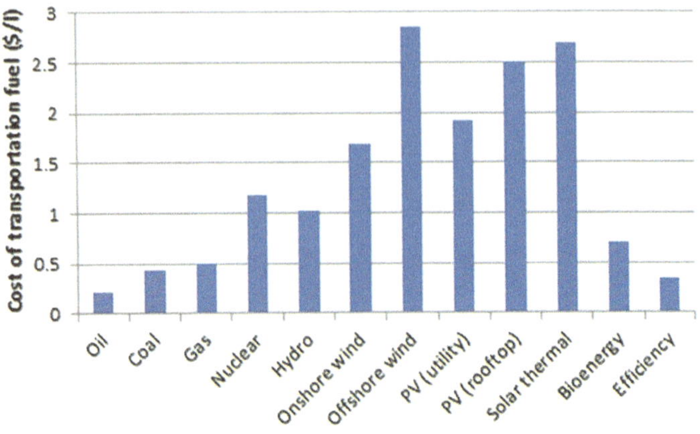

Fig. 7.21 Average global internalized costs of different transportation energy options (*Source* [64])

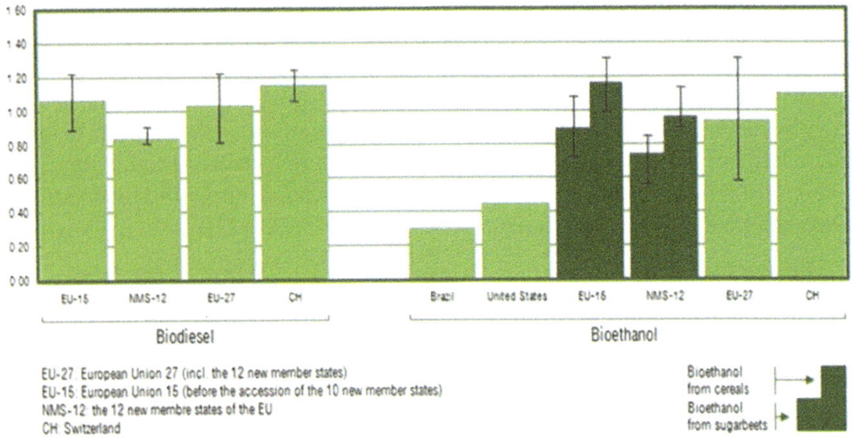

Fig. 7.22 Net production cost (in CHF/I) of biofuels according to the producing country/region (*Source* [65])

is on average three times as high than that in Brazil and twice as high than that in the USA, importing biofuel from these countries is not as simple or cost-effective as it may seem. In Fig. 7.22, production costs are given free-at-plant. Other costs that are added along the way are transport costs which are high because ethanol must first be transported by rail or truck from the production plant to the harbor, from where it will then be shipped to Europe by tanker, and finally distributed to end users. Throughout this trip, bioethanol will also need to be kept free of water to remain anhydrous. Lastly, customs duties and various tariffs may be added, thus increasing the final cost [65].

(g) **Risk of fires**

Dry biomass tends to be extremely flammable. Fire risks increase with the density of the crop and the low accessibility of the surrounding land, especially in dry parts of Europe with low rainfall. Potential fires will harm more than the energy crop in case of perennials placed near forests and shrub lands [66].

(h) **Stress on water resources**

The energy and the food crises are not the only ones threatening human survival. The water crisis is equally, if not more, important. Pressures on water supply are increasing worldwide due to population growth, migrations, global climate change, and intensive agriculture, among other things. Although it is predicted that by 2020 water use will have increased by 40%, it is highly doubted if this level of supply is even feasible. Instead, almost two-thirds of the world's population are expected to be under water stress conditions by 2025. Given the fact that irrigation increases yields of most crops by 100–400%, it is not surprising that water use has been growing at more than twice the rate of population growth over the last century, according to FAO [26, 57].

Biofuel production may impose further stress on water resources depending on crop type, irrigation demands, and location. Moreover, water is an essential part of the conversion process, thus increasing water demands even further. It is estimated that biorefineries consume 4 gallons of process water per gallon of bioethanol produced, largely from evaporative losses during the distillation of ethanol following fermentation [67, 68].

The EU target to replace 10% of transport fuels by renewables by 2020 requires additional water. However, different modes of transport present different levels of water efficiency. For example, trains or cars using bioelectricity are more water efficient than transport by car or airplane using bioethanol. Biofuel-based freight transport is most water efficient by ship or train; airplanes are least efficient [69].

(i) **Spread of genetically modified organisms**

In order to increase yields and environmental benefits of energy crops and to improve economic and energy efficiency while reducing agricultural inputs, genetic improvement has played an important role in the development of the biofuel industry. In the USA, Canada, and some South American countries such as Brazil and Argentina, planting of GMOs is already widespread. But in the EU, expansion of GMOs is heavily regulated and only 20 GM varieties—mostly corn and soya—have been approved for planting. However, genetically modified organisms (GMOs) are a controversial subject, and concerns have been raised regarding food safety and impacts on biodiversity, especially since, unlike food crops, genetically modified energy crops are not subject to the same rules for labeling [3].

There is considerable concern that biofuel development will lead to a wider spread of GMOs, because many energy crops have high invasive potential as they are fast growing, have many seeds, and regenerate easily [70].

7.1.3 Opportunities

(a) **Market growth perspectives**

Over the last decades, biomass and biofuel projects have been developed worldwide creating a large economic sector with significant business opportunities. They are considered strong market players with the potential to increase farming revenues, expand the productive capacity of existing cropland, and generate new jobs [71]. They also offer opportunities for investors to receive raw materials and to take advantage of the market price of the energy produced [72]. Although only 5% of biofuels is currently traded internationally, it is expected to rise in the future, and after 2020, lignocellulosic feedstock trade is likely to grow rapidly and supply biofuel plants of second- and third-generation biofuels [55].

Furthermore, production capacity is expected to rise as suggested by the establishment of many new projects around the world as EU member and other countries have set targets for biofuels share in their energy systems and issue directives to promote greater use of biofuels. Therefore, forecasts for the future of this market are very optimistic [3].

In Greece, the agricultural sector consists more than 5% of gross domestic product (GDP), almost three times the average 1.8% of the European Union (EU). Therefore, the companies involved in biomass and biofuels will find abundant sources of raw materials. Moreover, the commitment of the Greek government to replace 10% of current fossil fuels with biofuels by 2020 implies significant opportunities for the future [72].

(b) **Energy and climate change priority on policy agenda**

Countries across the globe are focusing on renewable energy to reduce greenhouse gas emissions to minimize climate change [73]. In general, even though biofuels are not deemed to provide a final solution to global warming, guidelines and directives in many countries are promoting biofuel use as a means to GHG reduction [3].

The United States Energy Independence Security Act of 2007 mandated the production of 36 billion gallons of biofuels by 2022, of which 21 billion gallons must be advanced biofuels [74]. The European Union has committed to at least 40% domestic reduction in greenhouse gas emissions compared to 1990 [75]. In 2009, the EU adopted the Renewable Energy Directive (RED), which sets a target of 10% of renewable energy in total transport fuel consumption by 2020. Biofuels and bioliquids are instrumental in helping EU countries meet this target [76]. Both the RED and the Fuel Quality Directive (FQD) also establish environmental sustainability criteria that biofuels consumed in the EU have to comply with. For example, biofuels should contribute to a reduction of at least 35% of greenhouse gas emissions in order to be taken into account [77].

However, not all biofuel types have the same efficiency. The need to identify biofuels with lower GHG emissions has led to initiatives to put forward biofuel certification according to carbon intensity. Figure 7.23 shows estimates of GHG reduction for different types of biofuels [3].

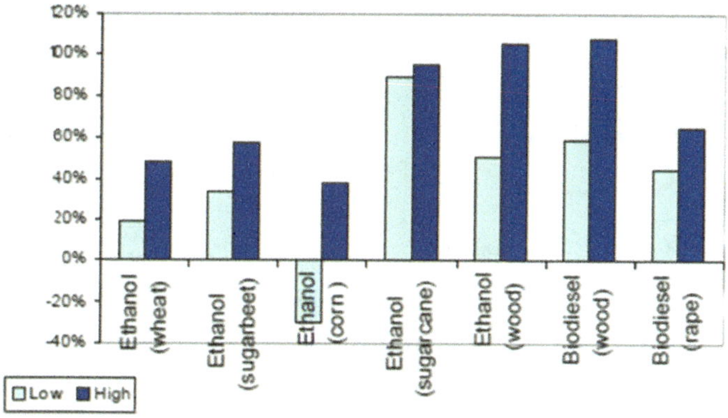

Fig. 7.23 GHG reductions for different biofuels (*Source* [3])

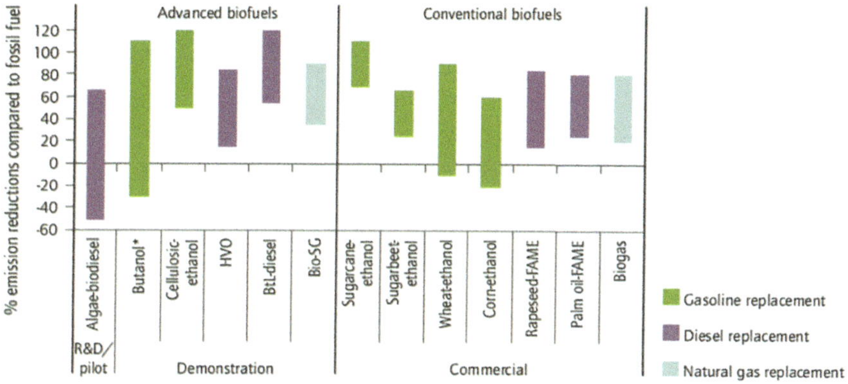

Fig. 7.24 Biofuel production and GHG emissions (IEA 2012) (*Source* [118])

Figure 7.24 shows percentage emission reductions of different biofuel production pathways, according to IEA (International Energy Agency). However, the significant differences between types of biofuels may be partly attributed to different methodologies used for assessing emissions [78].

(c) **Utilization of marginal lands**

Marginal lands are usually excluded from agricultural production due to their low productivity and profitability and their erosion susceptibility. However, they may be practically suitable to either grow short-rotation grasses or perennial vegetation, native to these lands. This way they will not only provide cellulosic biomass without competition with food crops, but also help to reclaim those lands along with substantial mitigation of greenhouse gases (GHGs) without posing any risk of food security. By using marginal, rather than primeland farmers may increase their income. Furthermore, by choosing the proper energy crop, they can potentially produce biofuel and at the same time enhance ecosystem services. The few available studies have

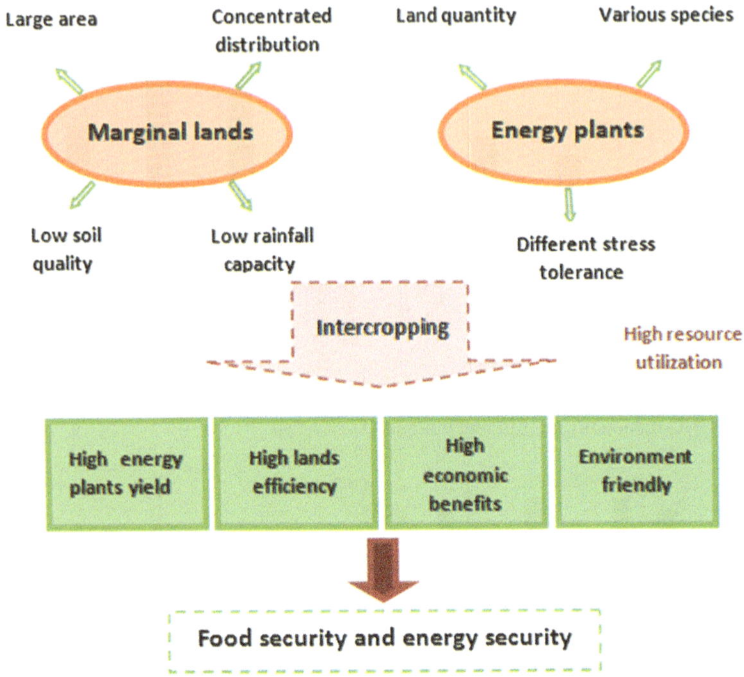

Fig. 7.25 Model chart of improving energy plant production through intercropping in marginal land to ensure food and energy security (*Source* [80])

strongly indicated that perennial vegetation can improve the soil and benefit the environment and biodiversity [79].

Figure 7.25 illustrates how the use of marginal land for biofuels can provide energy resources, improve the land output, and bring economic benefits, while preventing soil erosion, wind erosion, and desertification [80].

(d) Economically competitive in relation to fossil fuels

During the initial stages of their development, biofuels relied heavily on subsidies in order to make their price similar to fossil fuel-derived products [14, 81]. However, although in the past renewables were much more expensive than fossil fuels, recent developments have brought about significant changes. Since 2000, along with relevant increases in wind power and solar power, biofuel ethanol production has more than doubled, and biodiesel production has expanded nearly fourfold. These dynamic growth rates are driving down costs and evoking rapid advances in technologies. Consequently, new biofuel technologies are, or soon will be, economically competitive in relation to fossil fuels [82, 83].

7.1.4 *Threats*

(a) **Management practices for Sustainable Agriculture**

In order to meet the constantly growing energy demands and in an effort to become cost-effective, biofuel production is in danger of losing the sustainability scope and the environmental sensitivity which is necessary for the preservation of the planet. Inappropriate agricultural practices, large monocultures, unsuitable crop choices, and ineffective management and monitoring can lead to loss of sustainability and adverse results concerning pollution and the environment [84].

(b) **Higher risk of monoculture**

With the large scale of biofuel production comes a serious threat not only to the crop itself but also to the environment and to society. Firstly, monoculture may impose a long-term risk of crop failure because the crop is more vulnerable to pathogens and parasites and is in higher danger of crop extinction [85]. Also, the cycle of carbon, nitrogen, water, and soil fertility relies on crop and plant variety which is lost in the case of monocultures [86]. Moreover, large monocultures endanger biodiversity and distort landscape [13].

Finally, intensive monocultures may further distort land relations and increase land concentration to the detriment of small landowners. The expansion of energy crops as monocultures may displace thousands of farmers from their places and create extreme centralization of land [81]. The example of Brazil is a clear instance of smallholders who were forced to sell or leave their lands. The result is that 46% of the land in Brazil is concentrated in 1% of the rural population [86, 58].

(c) **Increase in food prices**

Food is probably the most basic and essential commodity, and its price fluctuations are much more than simple statistics; they might mean the difference between starvation and survival for certain parts of the planet. Since the beginning of the twenty-first century, the following decades of declining prices, basic food prices like wheat and rice have been steadily rising. During 2007–2008, export prices of wheat presented a staggering increase of 130%, 98% for rice, and 38% for corn, which was aptly reported as a global food crisis. Among the most gravely affected are the poor who spend 50–60% of their income on food (Fig. 7.26) [26, 87].

After almost two decades of rising food prices, the International Food Policy Research Institute (IFPRI) projects maize prices to rise 41% by 2020, with similar increases for oilseeds [2].

A more specific consequence of the food price increase is that it will particularly affect the poorer parts of the population. Urban consumers, especially in poor countries, are expected to suffer as food prices increase [88]. Especially since the prices of nearly all foods rise, households are asked to devote more and more parts of their income for sustainance. If this remains the trend, some nutrition studies show that the number of food-insecure people in the world would rise by more than 16 million for every percentage increase in the real prices of staple foods. This would mean that

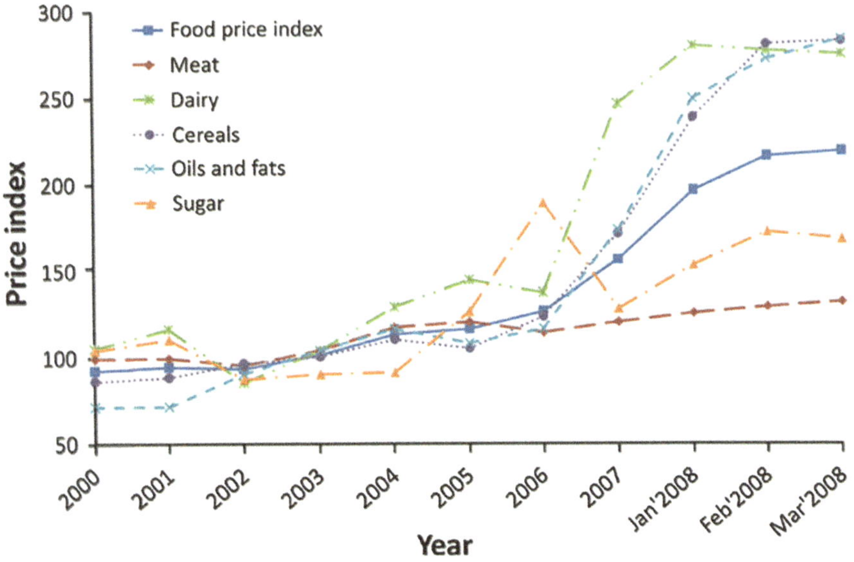

Fig. 7.26 Food commodity price indices from 2000 to 2007, and the first three months of 2008 (*Source* [26])

1.2 billion people could be chronically hungry by 2025—double than previously predicted [2].

Since 2000 global biofuel consumption has grown exponentially, particularly in Brazil and the USA. In Fig. 7.27, global biofuel consumption is presented together with the food price index [89].

Although food prices are influenced by a multitude of factors such as weather, stock levels, export policy changes, or even speculation, recent biofuel production has been repeatedly linked with the food price index [2]. The diversion of crops or croplands to biofuel production has been blamed for the tightening food supplies and rising food prices [26, 90]. Farmers are able to derive a greater profit by switching to biofuels [91, 92]. The "food versus fuel" debate has dominated the discussions, not without cause. It has been reported that the amount of crop needed to fill the tank of a sports utility vehicle could feed a person for an entire year [93].

The results of the SWOT analysis, regarding energy crops for biofuel production, are summarized in Table 7.4.

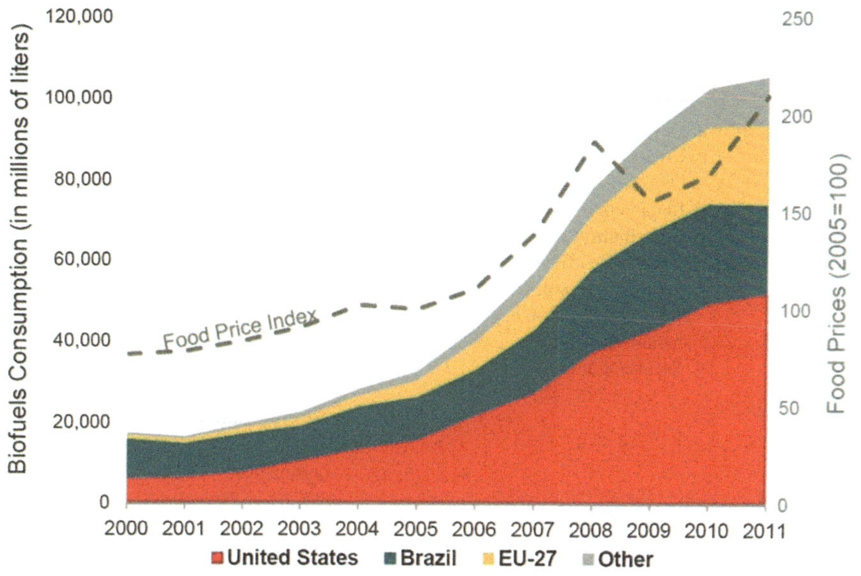

Fig. 7.27 Biofuel consumption and global food prices (*Source* [89])

7.2 Energy Crops for Food—SWOT Analysis

7.2.1 *Strengths*

(a) **Food security**

Although environmental sustainability and energy adequacy are major challenges for humanity, the most crucial issue is undoubtedly food security [94]. The challenge of feeding the constantly increasing world population on the same finite arable land of the planet is a difficult one. Land is a subject of competition between urban uses, edible crop agriculture, biofuel cultivation, and forestry. Moreover, in order to maintain biodiversity and climate stability and to preserve ecosystems, agricultural land cannot be expanded. The commercially traded edible crops are barely sufficient to meet food demand of the present world population. Certain regions, for example EU-27, do not even have a sufficient indigenous crop production [95].

Food insecurity is being monitored and analyzed by institutions like the International Food Policy Research Institute (IFPRI) [96]. The Food Security Risk Index has been developed to identify countries that may be affected by food shortages or even famine. Global food insecurity for 2012 is depicted in Fig. 7.28 [56].

Using energy crops for food helps to mitigate the food crisis and to take important steps to securing "the right to adequate food for all" as described by the Committee on World Food Security (CFS). According to the CFS, biofuel production absorbs such a large part of food crops that it poses a threat to food security [97]. All of

Table 7.4 Energy crops for biofuel production—SWOT analysis

Environment	Positive	Negative
Internal	Strengths	Weaknesses
	a. Development of rural areas. b. Creation of direct and indirect jobs c. Diversity of energy supply d. Independence from fossil fuel e. Contribution to the reduction of the greenhouse effect f. Reduction of soil erosion during the replacement of energy fields by farmland g. Soil remediation	a. Effects on quality of soil, water, and biodiversity b. Dependence on land availability c. Effects of land-use change d. Disease and pest sensitivity during the production of primary source e. Farmers' lack of knowledge or potential reluctance to accept dedicated energy crops f. Production cost g. Risk of fires h. Stress on water resources i. Spread of genetically modified organisms
External	Opportunities	Threats
	a. Market growth perspectives b. Energy and climate change priority on policy agenda c. Utilization of marginal lands d. Economically competitive in relation to fossil fuels	a. Management practices for sustainable agriculture b. Higher risk of monoculture c. Increase in food prices

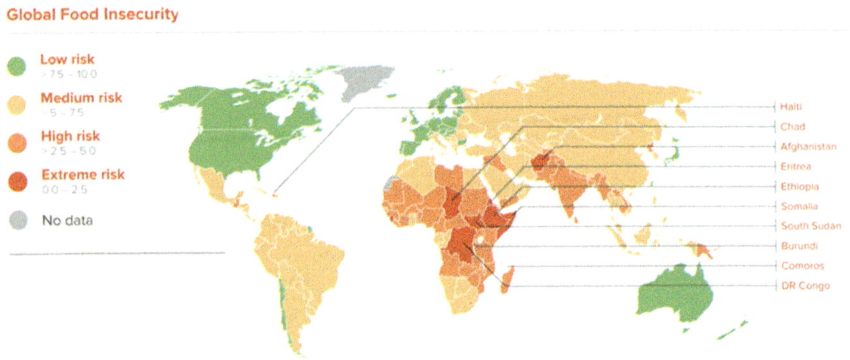

Fig. 7.28 Distribution of global food insecurity in 2012 (*Source* [56])

the chemical energy contained in 100% of all the world's crops in 2011 equaled just 14% of the world's primary energy consumption [98]. A particularly illustrative example of how increasing is this shift of food crops to biofuel production can be seen (Fig. 7.29). In the USA, ethanol consumes about 40% of corn and 25% of soy [99].

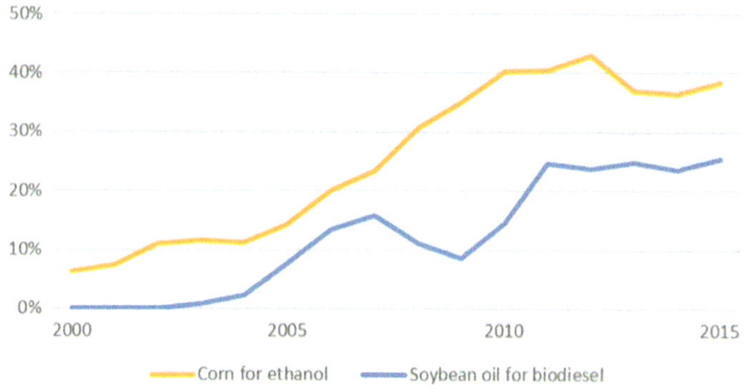

Fig. 7.29 Share of US corn used for ethanol and US soybean oil production used for biodiesel (*Source* [99])

Growing more food crops can aid food security but also presents alternatives for farmers. If the market is already saturated with food exports of a crop, this allows the crop to be diverted toward industrial use. This flexibility presents a number of advantages [100]:

- Farmers have more options for selling and therefore more economic security.
- Greater resource efficiency is beneficial to the environment.
- Flexible allocation of food crops in times of crisis ensures food security.
- Feed security is strengthened due to the high value of the protein-rich by-products of food crops.
- Increased global availability of food crops reduces the risk of shortages and speculation peaks and encourages market stability [100].

(b) **Non-Dependence on land availability**

Given the fact that feeding the world population will always rely on the proficient use of agricultural land for nutrition, it is clear that feed and food will always be the first priority of land use. The Food and Agriculture Organization (FAO) indicates that farmland currently totals 1550 million hectares throughout the world. Another 120 million hectares will be added—mostly in South America and in Africa—by 2030, but nothing afterward. Consequently, growing crops for food are an optimum use of available land without the threats that biofuel production imposes [101].

(c) **Avoidance of deforestation in order to find new land**

It is widely known that there is strong pressure for deforestation in order to find new lands for cultivation of energy crops and biofuel production. Deforestation is not only directly caused by biofuel establishment, but also has indirect effects, which influence entire regions or the world at large, and which will be increasingly important as large-scale biofuel development takes place. The latter concern is in part related to the additional GHG emissions from forest clearing, broader concerns about the loss of natural heritage and biodiversity, and the loss of environmental and other services and goods that forests provide to local communities. All of these consequences may be avoided by focusing agriculture on edible crops for food [102].

7.2.2 *Weaknesses*

(a) **Fewer opportunities for local economic development**

The low prices of some agricultural products have led in the recent years many farmers to poverty and abandonment of agricultural land.

In Greece in recent decades due to the declining productivity of (mono) crops, the increasing production costs, and modern trends toward imported agricultural products, the land under cultivation has dramatically been reduced to a staggering 6 million acres of abandoned farmland. One of the many examples of this reduction is the abandonment of the cereal cultivation because of almost zero producer profit with production costs exceeding 50 €/acre, which almost matches the average annuity of wheat in Greece (300 kg for 0.20 €/kg) [7].

Politicians and proponents stress the potential for biofuels to stimulate rural job creation and economic growth and increase energy independence as key reasons for providing public support for the industry. But in the rush to grow the sector, the benefits to rural communities may be muted or lost if federal, state, and local policies and programs that help determine the sector's ownership, scale, and structure do not sufficiently support rural development priorities [103].

The cultivation and sale of energy crops which are only used for human (or livestock) consumption involve the risk of failing to reverse the above-mentioned adverse situation and give lower opportunities for local economic development. Only with proper agricultural policy as crop restructuring (alternative crops) and incentives for young farmers, is there a prospect for local economic development.

(b) **Non-utilization of marginal lands**

Some of the plant species appropriate for production of biofuels are capable of growing in arid, water-stressed lands. Their cultivations also are less fertilizer and chemical intensive, which means that the farmers need to invest less capital on their farms and they use lands that are not suitable for other crops including food crops [23]. By cultivating only plant species for human consumption, possibilities of utilizing marginal lands are reduced. Marginal lands which are not suitable for the cultivation of edible crops are maximally utilized by the cultivation of energy plants.

7.2.3 *Opportunities*

(a) **Solution to the food crisis**

Focusing agriculture to edible crops in order to enhance food security is only viable if it is practiced based on sustainability principles. Intensive large-scale agriculture might have even worse repercussion on the environment than biofuels. An important opportunity for sustainable farming to feed to world would be to invest in small-scale producers, especially women who are practicing climate-resilient agriculture that requires few, if any, oil-based inputs. Investing in local food systems reduces the need for transportation fuel. Also, new agriculture technology and new plant species with enhanced features could increase the future food production without the need for more available land [104].

(b) **Second-generation biofuels and other renewable biofuel production**

First-generation biofuels made from food crops such as sugarcane, corn, wheat, and sugar beet require high agricultural inputs and increase food prices [105]. However, there are still options to avoid having to resort to fossil fuels.

Second-generation bioenergy crops and perennial energy crops in particular, which include a variety of native and non-native grasses and woody plants grown purely for energy production, present a more sustainable and affordable alternative with greater environmental benefits [106]. These require fewer inputs, have superior energy ratios, reduce GHG more than annual cropping, enhance water quality, and help preserve ecosystems. The lignocellulose in perennial forage crops is a more energy-dense material than the starch and sugars used from first-generation bioenergy crops. It represents a potentially vast and renewable source of biomass feedstock, and recent advances in technology enable production of biodiesel and renewable diesel from lignocellulose [44]. EU has issued the Renewable Energy Directive strongly promoting the development and use of the second- to third-generation biofuels, such as the bioethanol produced by lignocellulosic matrices and waste [107].

Moreover, second-generation biofuel technologies, such as thermochemical conversion of biomass, can be used to convert material such as crop residues into bioenergy [23, 108]. Biodiesel fuel can be made from new or used vegetable oils and animal fats, which are non-toxic, biodegradable, renewable resources [109, 110]. Third-generation biofuels (biofuel from algae) use non-edible raw material sources that can be used for biodiesel and bioethanol [111].

Finally, microbial products such as feed, fuel, food, industrial chemicals, and pharmaceutical products can be processed by employing a technology known as "solid-state fermentation" providing sustainable biofuel production and could be particularly beneficial to farmers in the developing world [112].

(c) **Development of new and non-edible varieties**

The possible contradiction between fuel and food production will be more easily solved by developing new and non-edible feedstock varieties [113].

Table 7.5 Non-edible energy crops, according to the EU most suitable for southern European countries and especially for Greece (*Source* [116])

Forest energy crops	Agricultural energy crops	
	Perennial crops	Annual crops
Two types of Eucalyptus (*Eucalyptus globules* Labill., *Eucalyptus camaldulensis* Dehnh.)	*Arundo donax* L.	*Hibiscus* L.
Robina pseudoacacia L.	*Miscanthus x giganteus GREEF et DEU*	
	Cynara cardunculus L.	
	Switchgrass (*Panicum virgatum* L.)	

While most grain crops are annuals, perennial grasses (switchgrass, miscanthus, giant reed, cardoon) grow continuously and are generally better able to survive environmental stresses such as drought. They also require less fertilizer and pesticide, making them ideal for cultivation on sub-optimal, underused, or abandoned land. The EU-funded "Optimization of Perennial Grasses for Biomass Production" (OPTIMA) project aims to use perennial grasses adapted to the Mediterranean environment as sources for biofuels and novel bioproducts. Perennial grasses are suitable for biomass production due to their high-yield potential, the high content of lignin, cellulose, and hemicelluloses polysaccharides, and their positive social and environmental benefits. Rural communities could generate new sources of income by growing these perennial bio-factories on marginal lands, without compromising existing crops grown for food. In addition, such grasses could potentially help reduce atmospheric greenhouse gasses by acting as carbon sinks [114]. Moreover, marginal lands and wastelands can be utilized with non-edible oilseed plants, which can provide raw material for biodiesel production without food competition and without disturbing indigenous biodiversity. For example, jatropha can be grown on degraded land, Chinese wild tallow tree on rough and waste sites, and rocket seed crop on arid land [115].

The non-edible energy crops, according to the European Union most suitable for southern European countries and especially for Greece, are divided as shown in Table 7.5 [116].

7.2.4 Threats

(a) **Dependence from fossil fuels**

Minimizing the use of edible crops in biofuel production may ensure food security; however, it increases dependence on fossil fuels to the detriment of the environment, energy security, and sustainability. Using alternatives to fossil fuels seems an imperative for the welfare of the planet and the human race, but if it is to be achieved without food risks, then biofuel production should be supplied with non-edible crops.

Table 7.6 Energy crops for food—SWOT analysis

Environment	Positive	Negative
Internal	Strengths	Weaknesses
	a. Food security b. Non-dependence on land availability c. Avoidance of deforestation in order to find new land	a. Fewer opportunities for local economic development b. Non-utilization of marginal lands
External	Opportunities	Threats
	a. Solution to the food crisis b. Second-generation biofuels and other renewable biofuel production c. Development of new and non- d. edible varieties	a. Dependence from fossil fuels b. Failure to comply with the world agenda priority for reversing the climate change

(b) **Failure to comply with the world agenda priority for reversing the climate change**

Many countries and especially EU have relied on biofuels to mitigate GHG emissions. By focusing energy crops for food or feed, countries are in danger of failing to meet the set targets and more importantly failing to reverse climate change. Since agriculture contributes about 14% of global annual greenhouse gas emissions, it is clear that energy crops that are not devoted to biofuels will further increase GHG emissions and diverge from the world agenda priorities regarding climate change [117].

Results of the SWOT analysis, regarding energy crops for food, are summarized in Table 7.6.

References

1. Thornley P (2002) Increasing biomass based power generation in the UK. Energy Policy 2006; 34:2087e99.http://www.sciencedirect.com/science/article/pii/S0301421505000637. Accessed 22 Nov 2016
2. Raswant V, Hart N, Romano M (2008) Biofuel expansion: challenges, risks and opportunities for rural poor people how the poor can benefit from this emerging opportunity. Paper prepared for the round table organized during the thirty-first session of IFAD's Governing Council, 14 February 2008. https://www.ifad.org/documents/10180/7eaa5557-b360-4928-b10d-6227ab93e2b8. Accessed 12 Dec 2016
3. Dufey A (2006) Biofuels production, trade and sustainable development: emerging issues. International Institute for Environment and Development, London. http://pubs.iied.org/pdfs/15504IIED.pdf. Accessed 12 Dec 2016
4. Carneiro P, Ferreira P (2012) The economic, environmental and strategic value of biomass Department of Production and Systems. University of Minho, Campus Azurem, 4800-058 Guimaraes, Portugal, 2012 Elsevier Ltd, Renewable Energy 44(0):18. http://www.energyplan.eu/wp-content/uploads/2013/09/Renewable-energy-scenarios-in-the-Portuguese-electricity-system-ICEE-Conference-2013.pdf. Accessed 15 Nov 2016
5. Energy crops for the production of liquid and solid biofuels in Greece, KAPE, CRES, 9. http://www.cres.gr/kape/pdf/download/energy_crops_2006_L.pdf. Accessed 22 Nov 2016

6. The Secretary-General's Advisory Group on Energy and Climate Change (AGECC) Energy for a sustainable future, Report and recommendations 28 April 2010 New York. http://www.un.org/millenniumgoals/pdf/AGECCsummaryreport[1].pdf. Accessed 10 Dec 2016
7. Danalatos N (2015) Innovative energy production from energy crops in Greece, Magazine "INNOVATION & DEVELOPMENT" printed version of www.ka-business.gr. http://www.ka-usiness.gr/pages/agricultural/8912/kainotomikh-paragwgh-energeias-apo-energeiakes-kalliergeies-sthn. Accessed 03 Dec 2016
8. Goldemberg J (2002) Brazilian energy initiative, world summit on sustainable development; Setembro, Joanesburgo, South Africa. http://www.solarthermalworld.org/sites/gstec/files/story/2015-06-20/goldembergsecondnewsletter.pdf. Accessed 22 Nov 2016
9. Plieninger T, Bens O, Hüttl RF (2006) Perspectives of bioenergy for agriculture and rural areas. Berlin-Brandenburg Academy of Sciences and Humanities, Brandenburg University of Technology, Germany, 35(2):125. http://www.researchgate.net/profile/Tobias_Plieninger/publication/233654571_Perspectives_of_bioenergy_for_agriculture_and_rural_areas/links/00b495210685d3ec55000000.pdf. Accessed 20 Nov 2016
10. The Bioenergy System Planners (BISYPLAN) Handbook (2012). http://bisyplan.bioenarea.eu/html-files-gr/01-02.html. Accessed 3 Dec 2016
11. Employ RES: Impact of Renewable Energy Policy on Economic Growth and Employment in the European Union (2006) https://ec.europa.eu/energy/sites/ener/files/documents/2009_employ_res_report.pdf. Accessed 10 Dec 2016
12. Piotrowski S, Carus M (2015) Nova-Institut EU-28 bioeconomy is worth 2 trillion euro, providing 19 million jobs. http://news.bio-based.eu/a-new-study-by-nova-institut-eu-28-bioeconomy-is-worth-2-trillion-euro-providing-19-million-jobs/. Accessed 12 Dec 2016
13. Sims RE (2003) Bioenergy to mitigate for climate change and meet the needs of society, the economy and the environment. Mitigation and Adaptation Strategies for Global Change 8:349. https://doi.org/10.1023/B:MITI.0000005614.51405.ce. https://link.springer.com/article/10.1023/B%3AMITI.0000005614.51405. Accessed 12 Dec 2016
14. Pimentel D (2008) Biofuels, solar and wind as renewable energy systems: benefits and risks. Springer Science & Business Media, Berlin, p 467. https://books.google.gr/books?id=YbxCAAAAQBAJ&dq=biofuel+Competition+fossil+fuels+renewable+sources.&hl=el. Accessed 22 Nov 2016
15. Kerebel C (2015) Renewable energy fact sheets of EU policy area: environment-energy. http://www.europarl.europa.eu/thinktank/el/document.html?reference=04A_FT%282013%29050704. Accessed 22 Nov 2016
16. Herzog VA, Lipman ET, Kammen MD (2001) Renewable Energy Sources, Energy and Resources Group Renewable and Appropriate Energy Laboratory (RAEL), University of California, Berkeley, USA, Encyclopedia of Life Support Systems (EOLSS) Forerunner Volume "Perspectives and Overview of Life Support Systems and Sustainable Development," Part 4C. Energy Resource Science and Technology Issues in Sustainable Development—Renewable Energy Sources. http://www.eolss.com. http://rael.berkeley.edu/old_drupal/sites/default/files/old-site-files/2001/Herzog-Lipman-Kammen-RenewableEnergy-2001.pdf. Accessed 22 Dec 2016
17. VIASPACE, Biomass versus fossil fuels, solar and wind. http://www.viaspace.com/biomass_versus_alternatives.php. Accessed 22 Dec 2016
18. Randall T (2015) Fossil fuels just lost the race against renewables-This is the beginning of the end, Bloomberg. http://www.bloomberg.com/news/articles/2015-04-14/fossil-fuels-just-lost-the-race-against-renewables. Accessed 21 Jan 2017
19. Eurostat Press Office, 30/2016—10 February 2016 Renewable energy in the EU Share of renewables in energy consumption in the EU rose further to 16% in 2014 Nine Member States already achieved their 2020 targets. http://ec.europa.eu/eurostat/documents/2995521/7155577/8-10022016-AP-EN.pdf/38bf822f-8adf-4e54-b9c6-87b342ead339. Accessed 22 Dec 2016
20. Independent from fossil fuels by 2050, Denmark-DK-The official web site of Denmark. http://denmark.dk/en/green-living/strategies-and-policies/independent-from-fossil-fuels-by-2050/. Accessed 22 Nov 2016

21. Hochman G, Rajagopal D, Zilberman D (2010) The effect of biofuels on crude oil markets, University of California-Berkeley, AgBioForum 13(2):3. http://www.agbioforum.org/v13n2/v13n2a03-hochman.htm. Accessed 23 Nov 2016
22. Miller C (2007) Bioenergy's "Top Five" List. BIOconversion Blog, Capitol Hill Club in Washington DC. http://bioconversion.blogspot.gr/2007_11_01_archive.html. Accessed 30 Jan 2017
23. Steer A, Hanson C (2015) Biofuels are not a green alternative to fossil fuels, The Guardian. http://www.theguardian.com/environment/2015/jan/29/biofuels-are-not-the-green-alternative-to-fossil-fuels-they-are-sold-as. Accessed 02 Dec 2016
24. Raboni M, Viotti P, Capodaglio GA (2015) A comprehensive analysis of the current and future role of biofuels for transport in the European Union (EU). Rev Ambient Água [online] 10(1):9–21. ISSN 1980-993X. http://dx.doi.org/10.4136/ambi-agua.1492. http://www.scielo.br/scielo.php?script=sci_arttext&pid=S1980-993X2015000100009. Accessed 22 Dec 2016
25. Biomass Boilers and Stoves (2014) Biomass boilers and the renewable heat incentive. http://www.biomassboilersandstoves.co.uk/#prettyPhoto. Accessed 22 Dec 2016
26. Koh LP, Ghazoul J (2008) Biofuels, biodiversity, and people: Understanding the conflicts and finding opportunities. Biol Conserv 141:2450–2460, Elsevier Ltd. http://www.sciencedirect.com/science/article/pii/S0006320708002954. Accessed 21 Jan 2017
27. International Energy Agency (2015) World energy outlook special report, energy climate and change. https://www.iea.org/publications/freepublications/publication/WEO2015SpecialReportonEnergyandClimateChange.pdf. Accessed 22 Dec 2016
28. European Union (2016) Perennial grasses better for biomass, Project reference: 289642 Funded under: FP7-KBBE, Country: Italy. http://cordis.europa.eu/result/rcn/153956_en.html. Accessed 12 Dec 2016
29. Cosentino LS, Scordia D (2011) Optimisation of perennial grasses for biomass production in mediterranean area, Optima (289642), Call: FP7 KBBE-2011-5. http://www.fibrafp7.net/portals/0/OPTIMA.pdf. Accessed 12 Dec 2016
30. SNIFFER ER05 (2010) Impacts of biomass and bioenergy crops on landscape, land use and the wider environment in Northern Ireland and Scotland, SNIFFER First Floor, Greenside House 25 Greenside Place Edinburgh EH1 3AA UK, Macaulay Scientific Consulting Ltd Craigiebuckler Aberdeen AB15 8QH UK. http://www.sniffer.org.uk/files/5013/4183/8003/ER05_Final_Project_Report_v3.1.pdf. Accessed 15 Dec 2016
31. Jankauskas B, Jankauskiene G (2003) Erosion-preventive crop rotations for landscape ecological stability in upland regions of Lithuania, Agricult Ecosyst Environ 95:129–142. http://www.sciencedirect.com/science/article/pii/S0167880902001007. Accessed 12 Dec 2016
32. Bical Energy, Miscanthus environmental profile, Bical Industrial Crops Ltd.; 2005. p. 10. https://www.yumpu.com/en/document/iew/42909839/miscanthus-environmental-profile-enagri. Accessed 15 Dec 2016
33. Pandey CV, Bajpai O, Singh N (2016) Energy crops in sustainable phytoremediation 54:58–73. http://www.sciencedirect.com/science/article/pii/S1364032115010485. Accessed 10 Dec 2016
34. Abhilash PC, Jamil S, Singh N (2009) Transgenic plants for enhanced biodegradation and phytoremediation of organic xenobiotics. Biotechnol Adv 27:474–488. https://www.ncbi.nlm.nih.gov/pubmed/19371778. Accessed 26 Jan 2017
35. Pandey VC, Singh K, Singh JS, Kumar A, Singh B, Singh RP (2012) *Jatropha curcas*: a potential biofuel plant for sustainable environmental development. Renew Sustain Energy Rev 16:2870–2883. http://www.jatropha.pro/PDF%20bestanden/Jatropha%20review.pdf. Accessed 26 Jan 2017
36. Ye W-L, Khan MA, McGrath SP, Zhao F-J (2011) Phytoremediation of arsenic contaminated paddy soils with *Pteris vittata* markedly reduces arsenic uptake by rice. Environ Pollut 159:3739–3743. https://www.ncbi.nlm.nih.gov/pubmed/21840633. Accessed 26 Jan 2017
37. Davis LC, Castro-Diaz S, Zhang Q, Erickson LE (2002) Benefits of vegetation for soils with organic contaminants. Crit Rev Plant Sci 21(5):457–491. Available at: http://www.tandfonline.com/doi/abs/10.1080/0735-260291044322. Accessed 26 Jan 2017

38. Passatore L, Rossetti S, Juwarkar AA, Massacci A (2014) Phytoremediation and bioremediation of polychlorinated biphenyls (PCBs): state of knowledge and research perspectives. J Hazard Mater 278:189–202. https://www.ncbi.nlm.nih.gov/pubmed/24976127. Accessed 26 Jan 2017
39. Fulekar MH, Singh A, Thorat V, Kausik CP, Eapen S. (2010) Phytoremediation of 137Cs from low level nuclear waste using Catharanthus roseus. Indian J Pure Appl Phys 48:516–519. http://nopr.niscair.res.in/bitstream/123456789/9912/1/IJPAP%2048(7)%20516-519.pdf. Accessed 26 Jan 2017
40. Cerne M, Smodis B, Strok M (2011) Uptake of radionuclides by a common reed (*Phragmites australis* (Cav.) Trin. Ex Steud.) grown in the vicinity of the former uranium mine at Zirovski Vrh. Nucl Eng Des 241:1282–1286. http://www.sciencedirect.com/science/article/pii/S0029549310002566. Accessed 26 Jan 2017
41. Van Ginneken L, Meers E, Guisson R, Ruttens A, Elst K, Tack MGT, Vangronsveld J, Diels L, Dejonghe W (2007) Phytoremediation for heavy metal-contaminated soils combined with bioenergy production. J Environ Eng Landscape Manag 15:4, 227–236. http://www.tandfonline.com/doi/pdf/10.1080/16486897.2007.9636935. Accessed 10 Dec 2016
42. Abreu CA, Coscione AR, Pires AM, Paz-Ferreiro J (2012) Phytoremediation of a soil contaminated by heavy metals and boron using castor oil plants and organic matter amendments. J Geochem Explor 123:3–7. https://www.researchgate.net/publication/257189070_Phytoremediation_of_a_soil_contaminated_by_heavy_metals_and_boron_using_castor_oil_plants_and_organic_matter_amendments. Accessed 26 Jan 2017
43. PandeyVC (2013) Suitability of *Ricinus communis* L. cultivation for phytoremediation of fly ash disposal sites. Ecol Eng 57:336–341. http://www.sciencedirect.com/science/article/pii/S0925857413001729?np=y. Accessed 26 Jan 2017
44. Littlejohn C, Curran T, Rainer W, Wratten S (2015) Farmland, food, and bioenergy crops need not compete for land. Solutions 6(3):34–48. http://www.thesolutionsjournal.com/node/237359. Accessed 12 Dec 2016
45. Rode M (2005) Energetische Nutzung von Biomasse und der Naturschutz, Natur und Landschaft (80):403–412. http://www.nul-online.de/Archiv/Archiv/Energetische-Nutzung-von-Biomasse-Reststoffen,QUlEPTQxMDQ5MjcmTUlEPTgyMDMw.html. Accessed 22 Nov 2016
46. Biomass Energy and Cellulosic Ethanol, Renewable Energy for America, Natural Resources Defence Council. http://www.nrdc.org/energy/renewables/biomass.asp. Accessed 23 Nov 2016
47. Renewable Fuel Standard: Potential Economic and Environmental Effects of U.S. Biofuel Policy (2011), Chapter: 5 Environmental effects and tradeoffs of biofuels. ISBN: 978-0-309-18751-0, https://doi.org/10.17226/13105. https://www.nap.edu/read/13105/chapter/7#206. Accessed 10 Dec 2016
48. BIOCORE: WP 7: Integrated assessment of overall sustainability, Final report on SWOT analysis and biomass competition (Deliverable D 7.7) (2013). http://www.biocore-urope.org/file/D7.7%20Final%20report%20on%20SWOT%20analysis%20and%20biomass%20competition%20.pdf. Accessed 20 Dec 2016
49. European Commission/Energy/Topics/Renewable energy/Biofuels/Land use change. http://ec.europa.eu/energy/en/topics/renewable-energy/biofuels/land-use-change. Accessed 22 Nov 2016
50. Casson A (1999) The Hesitant boom: Indonesia's oil palm subsector in an era of economic crisis and political change. Center for International Forestry Research, Bogor, Indonesia. http://www.cifor.cgiar.org/publications/pdf_files/CASSON.pdf. Accessed 21 Jan 2017
51. Weisenthal J (2011) Chart of the day: why The world has no choice but to buy more and more fertilizer. Dundee Securities. http://ehsmanager.blogspot.gr/2011/02/why-world-has-no-choice-but-to-buy-more.html. Accessed 28 Jan 2017
52. Hibbard K, Wilson T, Averyt K, Harriss R, Newmark R, Rose S, Shevliakova E, Tidwell V (2014) Energy, water, and land use. In: Melillo JM, Terese TC, Richmond, Yohe GW (eds) Climate change impacts in the United States: The Third National Climate Assessment,

U.S. Global Change Research Program, 257–281. https://doi.org/10.7930/j0jw8bsf. http://s3.amazonaws.com/nca2014/low/NCA3_Full_Report_10_Energy_Water_Land_LowRes.pdf?download=1, http://nca2014.globalchange.gov/report/sectors/energy-water-and-land. Accessed 28 Jan 2017
53. Go Gayathri G, Negri MC, Wang M, Wu M, Snyder S (2008) Use of marginal land and water to maximize biofuel production. In: Proceedings of the short rotation crops international conference; 2008 August 19–20; Bloomington, MN. Gen. Tech. Rep. NRS-P-31. Newtown Square, PA: U.S. Department of Agriculture, Forest Service, Northern Research Station: 15. https://www.nrs.fs.fed.us/pubs/gtr/gtr_nrs-p-31/15gopalakrishnan-p-31.pdf. Accessed 22 Nov 2016
54. Dufey A, Vermeulen S, Vorley B (2007) Biofuels: strategic choices for commodity dependent developing countries, International Institute for Environment and Development, London, UK, Common Fund for Commodities, Commodities Issues Series, 13. http://www.common-fund.org/fileadmin/user_upload/Publications/CFC_report_IIED_Biofuels_Strategic_Choices_2007.pdf. Accessed 24 Nov 2016
55. Popp J, Lakner Z, Harangi-Rákos M, Fári M (2014) The effect of bioenergy expansion: food, energy, and environment. Renew Sustain Energy Rev 32:559–578. http://www.sciencedirect.com/science/article/pii/S1364032114000677. Accessed 20 Dec 2016
56. The new climate economy Report 2014 Better growth, better climate: chapter 3: Land use. http://newclimateeconomy.report/2014/land-use. Accessed 28 Jan 2017
57. Ward S, Food vs. fuel the great ethical dilemma. School of Biosystems Engineering, University College Dublin. https://www.ucd.ie/foodandhealth/non-cms-content/2050/2050-pres-SWard.pdf. Accessed 27 Dec 2016
58. Gyulai I (2008) The biomass dilemma, CEEweb for Biodiversity and the National Society of Conservationists—Friends of the Earth Hungary. http://www.mtvsz.hu/dynamic/biomassen.pdf. Accessed 22 Dec 2016
59. Panoutsou C, Elbersen B, Böttcher H (2011) Energy crops in the European context, Biomass Futures project funded by the Intelligent Energy Europe Programme. http://www.biomassfutures.eu/public_docs/final_deliverables/WP8/D8.4%20Energy%20crops%20in%20the%20European%20context%20(contribution%20to%20FNR%20workshop).pdf. Accessed 20 Dec 2016
60. Stewart A, Cromey M (2011) Identifying disease threats and management practices for bio-energy crops. Curr Opin Environ Sustain 3(1–2):75–80. http://www.sciencedirect.com/science/article/pii/S1877343510001144. Accessed 20 Dec 2016
61. Ostwald M, Jonsson A, Wibeck V, Asplund T (2013) Mapping energy crop cultivation and identifying motivational factors among Swedish farmers. Biomass Bioenergy (50):25–34. Available at: http://liu.diva-portal.org/smash/get/diva2:575749/FULLTEXT01.pdf. Accessed 20 Dec 2016
62. Augustenborg CA, Finnan J, McBennett L, Connolly V, Priegnitz U, Müller C (2012) Farmers' perspectives for the development of a bioenergy industry in Ireland. Glob Change Biol Bioenergy 4:597–610. https://doi.org/10.1111/j.1757-1707.2011.01151.x. http://onlinelibrary.wiley.com/doi/10.1111/j.1757-1707.2011.01151.x/full. Accessed 27 Dec 2016
63. Growers Ready to Put Marginal Lands to Better Use, November 29, 2010, Posted in Seeding & Planting Source: Ceres, Inc., https://www.no-tillfarmer.com/articles/1290-growers-ready-to-put-marginal-lands-to-better-use. Accessed 27 Dec 2016
64. Schalk C (2015) Internalized costs results from the seeking consensus project TheEnergyCollective.com. http://www.theenergycollective.com/schalk-cloete/2221441/internalized-costs-results-seeking-consensus-study. Accessed 30 Jan 2017
65. Biodiesel Expedition 2014, Market Analysis. http://www.utm.my/biodiesel-expedition-2014/market-analysis/. Accessed 30 Jan 2017
66. European Environment Agency, Estimating the environmentally compatible bioenergy potential from agriculture. EEA Technical Report No 12/2007, ISSN 1725–2237. http://www.eea.europa.eu/publications/technical_report_2007_12/download. Accessed 30 Jan 2017

67. Pickett J, Anderson D, Bowles D, Bridgwater T, Jarvis P, Mortimer N, Poliakoff M, Woods J (2008) Sustainable biofuels: prospects and challenges. The Royal Society, London, UK. http://royalsociety.org/document.asp?id=7366. Accessed 21 Jan 2017
68. Pate R, Hightower M, Cameron C, Einfeld W (2007) Overview of energy-water interdependencies and the emerging energy demands on water resources. Report SAND 2007-1349C. Sandia National Laboratories, Los Alamos, New Mexico, USA, https://amfarid.scripts.mit.edu/resources/Media/Pate2007.pdf. Accessed 21 Jan 2017
69. Gerbens-Leenes W, Hoekstraa A (2011) The water footprint of biofuel-based transport, Energy Environ Sci 4:2658–2668. https://doi.org/10.1039/c1ee01187a. http://pubs.rsc.org/en/content/articlelanding/2011/ee/c1ee01187a#!divAbstract. Accessed 27 Dec 2016
70. Schubert R, Schellnhuber HJ, Buchmann N, Epiney A, Griesshammer R, Kulessa M, Messner D, Rahmstorf S, Schmid J (2010) Future bioenergy and sustainable land use. German Advisory Council on Global Change. http://www.wbgu.de/fileadmin/templates/dateien/veroeffentli81.chungen/hauptgutachten/jg2008/wbgu_jg2008_en.pdf. Accessed 27 Dec 2016
71. International energy agency: biofuels for transport—an international perspective. https://www.iea.org/newsroom/news/2004/may/2004-05-11-.html. Accessed 10 Dec 2016
72. Investing in the future—renewable sources of energy, Enterprise Greece Invest & Trade. http://www.investingreece.gov.gr/default.asp?pid=51&la=2. Accessed 22 Nov 2016
73. Somerville R C (2012) Energy Biosciences Institute, University of California, Berkeley, The food versus fuel debate, Biofuels 3(6):635–648. http://www.massey.ac.nz/~ychisti/BiofuelsInterview.pdf. Accessed 9 Feb 2017
74. Liu TT, McConkey BG, Ma ZY, Liu ZG, Li X, Cheng LL (2011) Strengths, weaknesses, opportunities and threats analysis of bioenergy production on marginal land Published by Elsevier Ltd. Selection and peer-review under responsibility of RIUDS, Energy Procedia 5:2381–2382.https://www.researchgate.net/profile/Tingting_Liu27/publication/241122933_Strengths_Weaknessness_Opportunities_and_Threats_Analysis_of_Bioenergy_Production_on_Marginal_Land/links/555bfead08aec5ac22329c7d.pdf. Accessed 15 Nov 2016
75. A Framework Strategy for a Resilient Energy Union with a Forward-Looking Climate Change Policy, Brussels, 25.2.2015 COM, 80 final. Energy Union Package Communication From The Commission to the European Parliament, The Council, The European Economic And Social Committee, The Committee of the Regions and the European Investment Bank 18 (2015). http://ec.europa.eu/priorities/energy-union/docs/energyunion_en.pdf. Accessed 22 Nov 2016
76. Directive 2009/28/EC of the European Parliament and of the Council of 23 April 2009 on the promotion of the use of energy from renewable sources and amending and subsequently repealing Directives 2001/77/EC and 2003/30/EC (Text with EEA relevance) http://eur-lex.europa.eu/legal-content/EN/ALL/?uri=CELEX%3A32009L0028. Accessed 27 Dec 2016
77. Directive 2009/28/EC of the European Parliament and of the Council of 23 April 2009 on the promotion of the use of energy from renewable sources and amending and subsequently repealing Directives 2001/77/EC and 2003/30/EC (OJ L 140 of 5.6.2009, pp. 16–62) http://europa.eu/legislation_summaries/energy/renewable_energy/en0009_en.htm. Accessed 22 Nov 2016
78. Ehlert P (2016) Why are fossil fuels cheaper than renewable energy? https://www.quora.com/Why-are-fossil-fuels-cheaper-than-renewable-energy. Accessed 27 Dec 2016
79. Muhammad AM, Umer R, Muhammad N, Shafaqat A, Athar H, Munazza G (2016) Biomass production for bioenergy using marginal lands, Sustainable production and consumption. http://dx.doi.org/10.1016/j.spc.2016.08.003. http://www.spc.ichemejournals.com/article/S2352-5509(16)30019-7/pdf. Accessed 27 Dec 2016
80. Zhang Z, Liu C, Zhang W et al (2015) Achieving food security and high production of bioenergy crops through intercropping with efficient resource use in China [J]. Front. Agr Sci Eng 2(2):134–143 http://engineering.cae.cn/fase/EN/10.15302/J-FASE-2015069#AbstractTab. Accessed 27 Dec 2016
81. Life cycle assessments applied to first generation biofuels used in France, Final report, February 2010, Direction Production et Energies Durables (DEPD)—ADEME. http://www.ademe.fr/sites/default/files/assets/documents/70548_final_report_lca_1st-generation-biofuels_france.pdf. Accessed 22 Nov 2016

82. Flavin C, Sawin JL, Mastny L, Aeck MH, Hunt S, Evitt AM, Stair P, Podesta J, Cohen AU, Hendricks B, Mohin T (2006) The renewable path to energy security. Worldwatch Institute, Center for American Progress, 6. http://www.worldwatch.org/renewables-becoming-cost-competitive-fossil-fuels-us. Accessed 25 Nov 2016
83. Noble J (2013) Renewable energy becomes cost competitive with fossil fuels. Sauder School of Business, Centre for Social Innovation & Impact Investing. http://www.sauder.ubc.ca/Faculty/Research_Centres/Centre_for_Social_Innovation_and_Impact_Investing/Programs/Clean_Capital/Clean_Capital_News_Archive_2013/Renewable_energy_becomes_cost_competitive_with_fossil_fuels. Accessed 10 Dec 2016
84. Dale BE, Anderson JE, Brown RC, Csonka S, Dale VH, Herwick G, Jackson RD, Jordan N, Kaffka S, Kline KL, Lynd LR, Malmstrom C, Ong RG, Richard TL, Taylor C, Wang MQ, Take a Closer Look: Biofuels Can Support Environmental, Economic and Social Goals, Environmental Science & Technology, 5. http://web.ornl.gov/sci/ees/cbes/Publications/Dale%20Bruce%20et%20al%20%202014%20EST%20Take%20a%20Closer%20Look%20-%20Proof.pdf. Accessed 24 Nov 2016
85. Weitzman LM (2000) Economic profitability versus ecological entropy. Quarterly Journal of Economics, February 2000, The Risk of Catastrophic Crop Failure, Economic Intuition, Summer 2001; based on. http://www.ncpa.org/sub/dpd/index.php?Article_ID=6882. Accessed 27 Dec 2016
86. Biofuels: when treatment becomes worse disease. http://biotechwatch.gr/agrofuels2. Accessed 24 Nov 2016
87. Von Braun J (2008) Rising food prices: what should be done? International Food Policy Research Institute, Washington DC, USA. http://www.ifpri.org/pubs/bp/bp001.asp. Accessed 21 Jan 2017
88. The Economist, The End Of Cheap Food (2007) http://www.economist.com/node/10252015. Accessed 22 Nov 2016
89. Kimberly E (2015) Biofuel policies: fuel versus food, forests, and climate change. CGD Policy Paper 51. Washington DC: Center for Global Development. http://www.cgdev.org/publication/biofuel-policies-fuel-versus-food-forests-and-climate. Accessed 30 Jan 2017
90. Rahman S H, Ahmed AM, Bauer A, Bhushan I, Carrasco B, Chatterjee S et al (2008) Soaring food prices: response to the crisis. Asian Development Bank, Manila, Philippines. http://www.adb.org/Documents/Papers/soaringfood-prices/soaring-food-prices.pdf. Accessed 21 Jan 2017
91. Swanepoel E (2007) Food versus fuel debate escalates, Engineering news. http://www.engineeringnews.co.za/article/food-versus-fuel-debate-escalates-2007-11-02. Accessed 22 Nov 2016
92. Brown L (2006) How food and fuel compete for land. Are increasing oil prices affecting agriculture's importance in the world market? Globalist. http://www.theglobalist.com/food-fuel-compete-land/. Accessed 22 Nov 2016
93. Byerlee D, De Janvry A, Sadoulet E, Townsend R, Klytchnikova I (2008) World development report 2008: agriculture for development. The World Bank, Washington DC, USA. http://siteresources.worldbank.org/INTWDR2008/Resources/WDR_00_book.pdf. Accessed 21 Jan 2017
94. Prusty BAK, Chandra R, Azeez PA, Biodiesel: freedom from dependence on fossil fuels? Gujarat Institute of Desert Ecology, Post Box # 83, Opp. Changleswar Temple, Mundra Road, Bhuj–370001, Gujarat, India, 2, 25. Retrieved http://precedings.nature.com/documents/2658/version/1/files/npre20082658-1.pdf. Accessed 02 Dec 2016
95. Kullander, S. AMBIO (2010) Food Security: Crops for People Not for Cars 39:249. https://doi.org/10.1007/s13280-010-0032-5. http://link.springer.com/article/10.1007/s13280-010-0032-5. Accessed 27 Dec 2016
96. Biofuels and food security, Balancing needs for food, feed, and fuel International Food Policy Research Institute (IFPRI) (2008) http://www.ifpri.org/publication/biofuels-and-food-security. Accessed 03 Dec 2016

97. F.A.O. (Food and Agriculture Organization), Biofuel development should not compromise food security, says CFS (2013), Rome. http://www.fao.org/news/story/en/item/202208/icode/ . Accessed 30 Nov 2016
98. Hanson C (2013) Food Security, Inclusive Growth, Sustainability, and the Post-2015 Development Agenda, Background Research Paper. http://www.post2015hlp.org/wp-ontent/uploads/2013/05/Hanson_Food-Security-Inclusive-Growth-Sustainability-and-the-Post-2015-Development-Agenda.pdf. Accessed 03 Dec 2016
99. Jeremy M (2016) Everything you ever wanted to know about biodiesel union of concerned scientists. http://blog.ucsusa.org/jeremy-martin/all-about-biodiesel. Accessed 30 Jan 2017
100. Carus M, Dammer L, Nova-Institut GmbH nova paper #2 on bio-based economy 2013-07, Food or non-food: Which agricultural feedstocks are best for industrial uses? http://www.ibbnetzwerk-gmbh.com/uploads/media/nova_institut_-food_or_non-food_paper.pdf. Accessed 08 Dec 2016
101. Movahedi N, Paillotin G, Neveu A, Will the planet have enough land to feed mankind in the 21st century? Académie d'Agriculture de France, momagri, movement for a world agricultural organization. http://www.momagri.org/UK/points-of-view/Will-the-planet-have-enough-land-to-feed-mankind-in-the-21st-century-_646.html. Accessed 02 Dec 2016
102. Gao Y, Skutsch M, Masera O, Pacheco P (2011) A global analysis of deforestation due to biofuel development—Working Paper 68, Center for International Forestry Research, 30.http://www.cifor.org/publications/pdf_files/WPapers/WP68Pacheco.pdf. Accessed 22 Nov 2016
103. Kleinschmit J (2007) Biofueling Rural development: making the case for linking biofuel production to rural revitalization, POLICY BRIEF NO. 5, Carsey Institute, University of New Hampshire, 1. http://www.iatp.org/files/258_2_96834.pdf. Accessed 02 Dec 2016
104. Fueling the food crisis: the cost to developing countries of US corn ethanol expansion, Action Aid International, USA Report, 7 (2012). http://www.ase.tufts.edu/gdae/Pubs/rp/ActionAid_Fueling_Food_Crisis.pdf. Accessed 02 Dec 2016
105. Evans G (2007) International Biofuels Strategy Project. Liquid Transport Biofuels—Technology Status Report, NNFCC 08-017, National Non-Food Crops Centre. http://www.globalbioenergy.org/uploads/media/0711_NNFCC_-__Liquid_Transport_Biofuels_Technology_Status_Report.pdf. Accessed 23 Nov 2016
106. Inderwildi OR, King DA (2009) Quo Vadis Biofuels. Energy Environ Sci 2:343. https://doi.org/10.1039/b822951c. http://pubs.rsc.org/en/Content/ArticleLanding/2009/EE/b822951c#!divAbstract. Accessed 22 Nov 2016
107. Council of the European Union. Proposal on indirect land-use change: Council reaches agreement. Luxembourg, 13th June 2014. http://www.consilium.europa.eu/uedocs/cms_data/docs/pressdata/en/trans/143191.pdf. Accessed 28 Jan 2017
108. Damartzis T, Zabaniotou A (2010) Thermochemical conversion of biomass to second generation biofuels through integrated process design—a review. Renew Sustain Energy Rev. https://doi.org/10.1016/j.rser.2010.08.003. http://www.sciencedirect.com/science/article/pii/S1364032110002558. Accessed 27 Dec 2016
109. Bozbas K (2008) Biodiesel as an alternative motor fuel: production and policies in the European Union. Renew Sustain Energy Rev 12(2):542–552 http://www.sciencedirect.com/science/article/pii/S1364032105000614. Accessed 27 Dec 2016
110. Wiesenthal T, Leduc G, Christidis P, Schade B, Pelkmans L, Govaerts L et al (2009) Biofuel support policies in Europe: lessons learnt for the long way ahead. Renew Sustain Energy Rev 13(4):789–800. http://www.sciencedirect.com/science/article/pii/S1364032108000166. Accessed 27 Dec 2016
111. HLPE (2013) Biofuels and food security. A report by the High-Level Panel of Experts on Food Security and Nutrition of the Committee on World Food Security, 13–14, Rome. http://www.fao.org/fileadmin/user_upload/hlpe/hlpe_documents/HLPE_Reports/HLPEReport-5_Biofuels_and_food_security.pdf. Accessed 22 Nov 2016
112. BioMed Central (2015) Renewable biofuel production avoids competition with food resources. Science Daily. www.sciencedaily.com/releases/2015/01/150129204153.htm. Accessed 3 Dec 2016

113. EIE-05-113, BIODIESEL CHAINS, Promoting favourable conditions to establish biodiesel market actions, Strategy formation, D 17 SWOT Analysis, Case studies: Belgium, Bulgaria, Cyprus, Greece, Poland, Romania, CRES—Center for Renewable Energy Sources, 19th km Marathonos Avenue, 190 09 Pikermi, Greece, 7, 27–32. http://www.cres.gr/biodiesel/downloadsreports/Other/SWOT_Oct_07.pdf. Accessed 15 Nov 2016
114. Cosentino SL, Scordia D, Testa G, Copani V, Alexopoulou E, Christou M, Monti A, Zegada-Lizarazu W, Fernando AL, Duarte P, Curt MD, Panoutsou C, Soldatos P, Porqueddu C, Gulias J, Nogues S, Robson P, Webster R, Osborne B, Sharma S, Arora K, Yan J, Pari L, Reinhardt G, Mueller-Lindenlauf M, Mentaberry A, Van den Berg D, Spapperi S, Varga SS, De Jamblinne P (2012) OPTIMA—Optimization of perennial grasses for biomass production in the Mediterranean area- In: 20th European Biomass Conference and Exhibition. Milan, Italy. http://www.fibrafp7.net/portals/0/danilo_scordia_day_2.pdf. Accessed 03 Dec 2016
115. Ahmad M, Teong LK, Sultana S, Zafar M, Biodiesel production from non food crops: a step towards self reliance in energy. School of Chemical Engineering, Universiti Sains Malaysia, Nibong Tebal-14300, Penang, Malaysia. http://www.formatex.info/energymaterialsbook/book/239-243.pdf. Accessed 27 Dec 2016
116. Boukis I, Vassilakos N, Kontopoulos G, Karellas S (2009) Policy plan for the use of biomass and biofuels in Greece, Part I: available biomass and methodology. Renew Sustain Energy Rev 13(5):977. www.elsevier.com/locate/rser. Accessed 28 Jan 2015
117. Turral H, Burke J, Faurès JM (2011) Climate change, water and food security. FAO (Food and Agriculture Organization) Land and Water Division, Food and Agriculture Organization of the United Nations, Rome, xxiii. http://www.fao.org/3/a-i2096e.pdf_. Accessed 03 Dec 2016
118. Diop D, Blanco M, Flammini A, Schlaifer M, Kropiwnicka MA, Markhof MM (2013) Assessing the impact of biofuels production on developing countries from the point of view of Policy Coherence for Development, Contract N° 2012/299193, FWC COM 2011 - Lot 1—Studies and Technical Assistance in all, Sectors, Final report, February 2013. https://ec.europa.eu/europeaid/sites/devco/files/study-impact-assesment-biofuels-production-on-development-pcd-201302_en_2.pdf. Accessed 12 Nov 2016
119. Papazoglou GE (2016) Use of energy crops for phytoremediation purposes, SPEED, London, UK. http://www.core-community.net/wp-content/uploads/2016/04/Use-of-energy-crops-for-phytoremediation.pdf. Accessed 26 Jan 2017

Chapter 8
Alternative Strategies

SWOT Analysis. Alternative strategies are then based on particular combinations of the four strategic factors of both options, i.e., "growing energy crops for biofuels or for food."

SO Alternative strategies aim at using the strengths to take advantage of the opportunities.

ST Alternative strategies attempt to use the strengths as a way to avoid threats in the environment.

WO Alternative strategies try to overcome weaknesses by taking advantage of opportunities.

WT Alternative strategies suggest defensive actions to minimize weaknesses and avoid threats [1].

The sets of alternative strategies for the two studied options are further discussed and analyzed below and summarized in the two SWOT matrix/alternative strategies tables (Tables 8.1 and 8.2).

- **Energy crops for biofuels: Alternative strategies** (Table 8.1).

SO Alternatives to Bioenergy

Other renewable energy sources can complement bioenergy [2]. Dedicated land can also be used for alternatives to bioenergy like solar photovoltaic (PV) cells which convert sunlight directly into energy. PV systems are more efficient and use less water. For example, they can generate more than 100 times the usable energy per hectare than bioenergy. Because electric motors can be two to three times more efficient than internal combustion engines, solar PV can result in much more usable energy per hectare for vehicle transport compared to bioenergy [3].

SO Increasing Yields of Biofuels Technologies

By employing technologies that maximize yields in terms of productivity per ton of biomass, more energy can be produced using less feedstock [4].

A. Paschalidou et al., *Using Energy Crops for Biofuels or Food: the Choice*, Green Energy and Technology, https://doi.org/10.1007/978-3-319-63943-7_8

Table 8.1 Energy crops for biofuels: SWOT matrix/alternative strategies

Internal Factors / External Factors	Strengths (S)	Weaknesses (W)	
	a. Development of rural areas. b. Creation of direct and indirect jobs c. Diversity of energy supply d. Independence from fossil fuel e. Contribution to the reduction of the greenhouse effect f. Reduction of soil erosion during the replacement of energy fields by farmland g. Soil remediation	a. Effects on quality of soil, water and biodiversity b. Dependence on land availability c. Effects of land use change d. Disease and Pest sensitivity during the production of primary source e. Farmers' lack of knowledge or potential reluctance to accept dedicated energy crops f. Production cost g. Risk of fires h. Stress on water resources i. Spread of genetically modified organisms	
Opportunities (O) a. Market growth perspectives b. Energy and climate change priority on policy agenda c. Utilization of marginal lands d. Economically competitive in relation to fossil fuels	**SO Alternative strategies** • Alternatives to bioenergy • Increasing yields of biofuels technologies • Feedstock production from energy crops on marginal lands • Perennial energy crops integrated in agroecosystems	**WO Alternative strategies** • Use of energy crops as Shelterbelts • Employing water saving practices • Energy plants with super traits and super yields • Utilizing Waste Lands • Improving results from photosynthesis • Tackling consequences for biodiversity and conservation • Improving biomass yield • Research into lignin content and composition	**OT Alternative strategies** • Feedstock diversification • Governmental policies regarding biofuels production
Threats (T) a. Management practices for Sustainable Agriculture b. Higher risk of monoculture c. Increase in food prices	**ST Alternative strategies** • Policies for smallholders to participate in biofuel development • Combining species together	**WT Alternative strategies** • Policies promoting sustainable farming activities to avoid environmental risks • Mitigation of the effects of natural risks • Effective management of disease threats	
	SW Alternative strategies • Development of process technologies for next generation feedstocks		Synthesis

SO Feedstock Production from Energy Crops on Marginal Lands

Growing dedicated energy crops on marginal lands provides biomass together with a number of benefits for the local ecosystems. The amount of biomass produced and the extend of environmental benefits vary with the type of marginal land as well as management. Agricultural policy offering site-specific management strategies and guidelines for feedstock production from energy crops on marginal lands enables farmers to ensure that the plants selected for production in remote areas have the productivity to be competitive as well as beneficial. Such a strategy would be particularly suitable for developing countries, many of which are both net food and fuel importers [5, 6].

Table 8.2 Energy crops for food: SWOT matrix/alternative strategies

Internal Factors / External Factors	**Strengths (S)** a. Food security b. Non-Dependence on land availability c. Avoidance of deforestation in order to find new land	**Weaknesses (W)** a. Fewer opportunities for local economic development b. Non-utilization of marginal lands	
Opportunities (O) a. Solution to the food crisis b. Second generation biofuels and other renewable biofuel production c. Development of new and non-edible varieties	**SO Alternative strategies** • Non-food crops • Building better biofuel crops/new generation of plants • Sufficient land resources to feed the world's population	**WO Alternative strategies** • Gender-differentiated risks and promotion of gender equality	**OT Alternative strategies** • Biotechnology development
Threats (T) a. Dependence from fossil fuels b. Failure to comply with the world agenda priority for reversing the climate change	**ST Alternative strategies** • Energy saving policies	**WT Alternative strategies** • Diversification energy used in transport	
	SW Alternative strategies • Deforestation and second generation biofuels		Synthesis

SO Perennial Energy Crops Integrated in Agroecosystems

The introduction of perennial energy crops into annual food crop systems can achieve the desired combination of energy crops grown both for food and for biofuels. They can be integrated in agroecosystems in both space and time thus providing a range of ecosystem services that enhance soil, water, and biodiversity. Perennials can be on steep slopes and streamside buffers to generate more reliable yields in years of drought and flood while at the same time prevent soil erosion and benefit water quality. They can also be incorporated in annual crop rotations to improve soil properties such as water and nutrient holding capacity. By setting out incentives for these synergies between food and fuel, sustainable agricultural systems are created [7]

ST Policies for Smallholders to Participate in Biofuel Development

To minimize the prospect of land grabbing which may be brought about by rapid biofuel production growth but also to ensure that smallholders are included in the relevant economic development policy measures and pro-poor strategies should be designed and implemented. This would require public–private sector partnerships, providing farmers with access to supply chains, transport, and processing plants. Furthermore, to achieve the necessary economies of scale, existing institution like cooperatives or producer companies can ensure the interests of the poor by accumulating and attracting capital and partnerships for the necessary investments and organizing feedstock supplies in large quantities. Finally, smallholder farmers and rural people often lack legal recourse and should be protected from unfair business practices [8, 9].

ST Combining Species Together

In order to avoid large monocultures and the ensuing dangers and in an effort to imitate natural grasslands, combining different species together is a solution that could provide the required feedstock but also provide habitats for more diverse populations of organisms. Indeed, some research has suggested that natural grasslands have higher productivity than pure stands of grassland species. However, the use of diverse genotypes may limit yield in older varieties of some crops, such as maize and wheat [10].

WO Use of Energy Crops as Shelterbelts

To avoid replacing food with fuel, bioenergy crops can be integrated into the present farming system as shelterbelts, taking full advantage of their associated benefits. This way, apart from securing food production, energy crops can enhance ES delivery from the farming system, improving its sustainability and reducing its external costs. Improved food yield due to shelter-enhanced production and energy crop yields can counterbalance the extra costs [11].

WO Employing Water-Saving Practices

Choosing the right feedstock can contribute to reducing water demands. Some feedstocks, such as sugar cane, require considerable quantities of water, while others such as jatropha require less. In dry areas, the issue could be addressed by investing in soil management and water-saving technologies, some of which are not cost-efficient at present. Improvement in crop productivity as well as shifting from high water-use biofuel crops to drought-tolerant crops are also options worth considering [9].

WO Energy plants with Super Traits and Super Yields

Apart from choosing suitable crops and their residues, another potential option is to increase productivity. With advancements in genetic engineering, there are grains with traits that confer drought tolerance, heat tolerance, pest tolerance or pesticide tolerance. A system that would offer farmers improved genetics and agronomic practices for specific bioenergy crops, utilizing new integrated biorefineries, may enhance productivity, reduce input demands, and be more environmentally friendly [4].

WO Utilizing Wastelands

Saline or wastelands and abandoned farmlands can be cultivated to avoid competition for land. By adding irrigation water, recirculating water systems or using saline ponds that grow algae in the desert some degraded or dry lands might produce biofuels while avoiding this competition with food and carbon storage. Many companies are researching traits such as salt tolerance that will open up lands with previously unsuitable soils or water sources [4, 12].

WO Improving Results from Photosynthesis

Recent technologies funded in the ARPA-E PETRO project aim to create plants that capture more energy from sunlight and convert that energy directly into fuels. These technologies optimize the biochemical processes of energy capture and conversion to develop farm-ready crops that deliver more energy per acre with less processing. This may create biofuels for half their current cost, finally making them cost-competitive with fossil fuels [4].

WO Tackling Consequences for Biodiversity and Conservation

The development of more efficient biomass conversion pathways together with biotechnology has the potential to maximize yields and minimize land demand and input requirements. Lower land demands would contribute to fewer dangers biodiversity, natural habitats, and the environment in general. Although genetically engineered crops present a serious concern, genetic modifications can be limited to non-edible crops such as switchgrass [13].

WO Improving Biomass Yield

Maximizing biomass yields is essential to avoid land competition. This can be achieved by modifying plant architecture in a number of ways. The prevention of flowering would procure better yields from leaves, stalks, and roots and extend the period of biomass accumulation. Additionally, eliminating production of fruit or seeds is likely to increase total biomass and may also protect against unintentional spread of energy crop species. Selection for leaf retention can tackle problems like late-season leaf dropping which reduce yield. More generally, stem thickness, plant height, tiller number, and leaf density affect biomass productivity. Finally, to increase water- and nutrient-capture efficiency, root architecture may also be altered [10].

WO Research into Lignin Content and Composition

Another way to increase the production of biofuels is to grow feedstocks that are more easily converted to fuels. Research has shown that lignin content is directly related to cellulose and hemicellulose deposition and therefore to production yields [14]. Studies of variation in lignin content and composition have indicated that it is possible to significantly improve the digestibility of biomass by selecting for certain lignin compositions [15]. Consequently, there is a lot of interest in genetically altering the chemical composition or structure of biomass to render the conversion process more efficient and less expensive [10].

Other plant features that are worth considering altering are acetate release, which inhibits conversion and polysaccharides content in cell walls, which increases biofuel yield. By changing these aspects of biomass composition and engineering energy crops to produce more fermentable sugars, biofuel production becomes more cost-effective and less harmful [16, 17].

WT Policies Promoting Sustainable Farming Activities to Avoid Environmental Risks

Policies promoting sustainable farming activities can substitute intensive practices which can lead to soil degradation and depletion of natural resources. Sustainability principles can be observed by providing incentives to farmers to encourage them to mitigate the effect of climate change [9].

WT Mitigation of the Effects of Natural Risks

Effects of natural risks like droughts or floods can be mitigated through insurance mechanisms so that farmers are less vulnerable to them [9].

WT Effective Management of Disease Threats

It is important plan for and implement a range of crop management strategies to avoid disease or, if necessary, to provide a cure. The most valuable control strategy will be the use of resistant cultivars, supplemented with crop monitoring practices to detect diseases as soon as possible.

Natural disease suppression can be enhanced with the use of soil amendments, which together with crop hygiene practices will reduce the build up of pathogen inoculum. To deal with disease, targeted applications of microbial bioinoculants and/or fungicides are suggested. These will also deliver other crop benefits such as root growth promotion, drought tolerance, and increased oil production [18].

However, certain species such as switchgrass (*Panicum virgatum*) and *Miscanthus*, which are replanted every decade are threatened by a wide variety of viruses, fungi, insects, and nematodes. In this case, the best strategy will be diversification of the plantations to maximize the genetic diversity of relevant tolerance and resistance pathways [10].

OT Feedstock Diversification

Different feedstock options can minimize the use of edible energy crops for biofuels and offer various alternatives than reduce land competition, food insecurity, and GHG emissions. Corn stover, sugarcane bagasse, woods and forestry residues and energy grasses as well as new inedible oilseed crops such as jatropha, carinata, and camelina are notable choices [4]. Also, perennial grasses such as switchgrass, miscanthus (*Miscanthus* spp.), reed canary (*Phalaris arundinacea*), and giant reed (*Arundo donax*) are candidate grass species for cellulosic ethanol production especially on agriculturally degraded lands [11, 19]. High-diversity mixtures of grassland species can even provide greater bioenergy yields and GHG reductions than certain conventional bioethanol or biodiesel production systems [20].

Another potential feedstock is waste biomass which includes agricultural, forestry, and municipal wastes as well as timber processing wastes, urban waste wood, and landfill methane [21]. Horticultural waste biomass could also be a potential source of cellulosic feedstock [3, 19].

Producing biofuel from microalgae or macroalgae is another potential source of biofuel feedstock which has no land demands, does not threaten food production, and may provide further benefits of wastewater remediation or nutrient reduction [11]. Aquatic unicellular green algae, such as *Chlorella* spp., are considered for biodiesel production owing to their high growth rate, population density, and oil content [22]. Algae have much higher productivity (90,000 l of biodiesel per hectare [l/ha]) than soybean (450 l/ha), rapeseed (1200 l/ha), or oil palm (6000 l/ha) [2, 23].

OT Governmental Policies Regarding Biofuels Production

In the last decade, governmental policies have played a major role in biofuel development. By setting targets, providing subsidies, and granting incentives, they have encouraged biofuel production and consumption. Import tariffs have also played a significant role by protecting national industries from external competition. However, in light of recent research regarding biofuel disadvantages, governments are required to face the dual challenge of preserving the continued expansion of the biofuel sector while abiding to ecologically sustainable production requirements. A further concern will be to decrease the support that has so far sustained biofuel development in order to limit biofuels' burden on national budgets while at the same time ensuring that sustainability principles and climate change priorities are duly observed [24].

Biofuel production will continue and expand as long as government policies incorporate sustainability aspects: To limit biofuels based on grains, sugar, or vegetable oils, second-generation biofuel technologies and large-scale production facilities of second-generation biofuels should be developed. Also, requirements should be set for substantial reductions in GHG life-cycle emissions. Furthermore, to avoid the consequences of indirect land use change, feedstock origin certifications can be mandatory [12, 24].

In general, fuels with high greenhouse gas emissions and environmental impact can be avoided, if policy makers are willing to set environmental standards and to expand biofuel production while achieving climate protection objectives [25].

SW Development of Process Technologies for Next-Generation Feedstocks

Diversifying the biofuel feedstock cannot achieve the desired results if the processing technology does not follow. The key challenge is to take full advantage of the two primary converting technologies, biochemical and thermochemical conversion, and especially enzymatic hydrolysis, which is the key to cost-effective bioethanol production, according to some experts [21, 26]. A more effective thermochemical conversion pathway is gasification because all organic matter in biomass (including lignin) can be converted to liquid fuel [12].

- **Energy crops for food: Alternative strategies** (Table 8.2).

SO Non-Food Crops

Biofuel production from non-edible crops is a significant way to ensure food supply without waiving the chance to stabilizing farmers' income [27] . Species that are unsuitable for human or animal consumption but rich in energy content, like Miscanthus, Switchgrass, and Poplar could provide useful cash revenue to farmers and would complement food production rather than displace it [28, 29]. Other non-edible plants such as pongamia and jatropha are suitable for degraded lands, and tropical sugar beet can grow in alkaline and sodic soils [4, 9] (see Appendix Fig. A.3).

SO Building Better Biofuel Crops/New Generation of Plants

In order to maintain land availability for food, biofuel production could be redirected to lands that are unsuitable for food crops, like semiarid land, or prone to drought which account for 18% of Earth's terrestrial surface. This would require engineering high-biomass plants that are resistant to a variety of environmental stressors, such as drought, flooding, salt, cold, and pests [10]. Also, while waterlogged acres are typically difficult to farm with food crops, they provide an opportunity for maximal biomass production using perennials, especially if flood-tolerant species are developed. The United Nations Food and Agriculture Organization (FAO) estimates that almost 2% of irrigated lands are lost to salinization each year. Since salinized soils generally have low productivity and are not utilized for food crop production, salt-tolerant species, such as prairie cordgrass and *Eucalyptus* species, could be considered for biofuel production [10, 30].

Another feature that would extend the range of bioenergy crops that do not threaten food production is cold-tolerant crops hybrids such as Miscane, a cross between tropical sugarcane and cold-tolerant *Miscanthus* species. Miscane is frost and rhizomes freezing resistant and can become dormant during short days, allowing the plants to survive over cold periods like winter [10].

SO Sufficient Land Resources to Feed the World's Population

According to FAO studies, the total land area under crop production can reach up to 1645 M ha in 2050 (Fig. 8.1). It is further suggested there is still sufficient land resources to feed the world's population for the foreseeable future, and most countries are far from reaching the limits of their available land for agriculture. Moreover, there is still considerable room for yield improvements in developing countries, which will significantly decrease the land requirement for food and feeds [31].

ST Energy-Saving Policies

The dilemma "food or fuel" was caused by a combination of growing energy demands and growing concern of the environmental impact of fossil fuel use. Therefore, in order to devote energy crops and available land to food production, energy demands should be met with alternative solutions. An important step toward decreasing energy demands is energy saving policies and energy efficiency measures. Moreover, within the use of renewable energy sources the non-depletable sources (wind, sun) should be preferred to the depletable ones (biomass) and should be subjected to strict environmental and sustainability regulations [32].

WO Gender-Differentiated Risks and Promotion of Gender Equality

Food price increases disproportionally affect the poorer parts of the population, most often in developing countries, where women are more vulnerable than men to nutritional problems during times of food scarcity. Ensuring food security facilitates poor families' access to food and more particularly benefits women's food security [33] .

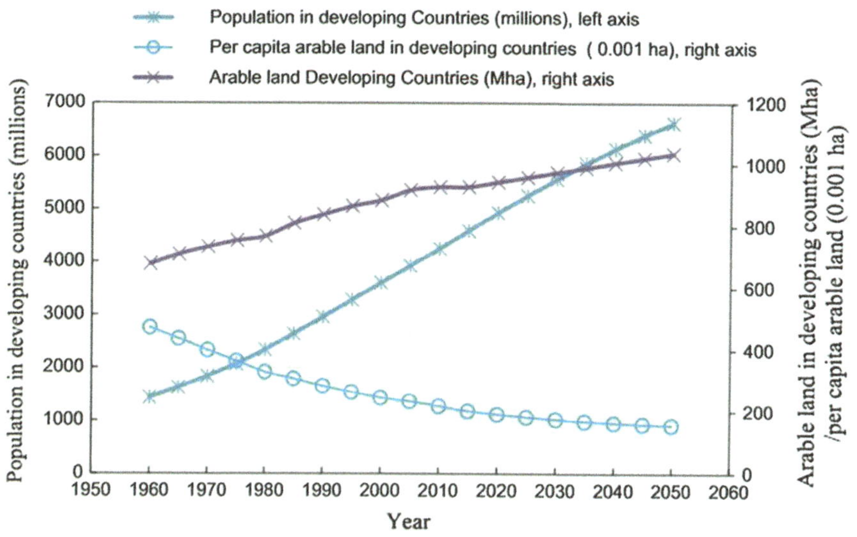

Fig. 8.1 Expected arable land expansion toward 2050 (*Source* [31])

WT Diversification Energy Used in Transport

The main antagonists of food production are first-generation biofuels. Acknowledging this threat to food security but also to the environment, EU has issued the "Transport 2050" plan, which specifies targets in terms of sustainable mobility to be achieved by 2050. Along these lines, second-generation biofuels, electric vehicles, and hydrogen cars present promising alternatives to both fossil fuels and first-generation biofuels [34].

OT Biotechnology Development

Land-use conflicts can be reduced by utilizing biotechnology to develop higher-yielding biofuel crops, more resistant to pests, diseases, or abiotic stresses [22, 23]. Furthermore, dedicated biofuel crops may be genetically modified to grow faster, have lower lignin content, in order to enhance the efficiency of cellulosic ethanol production [13]. Biotechnology can thus lead to increased supply and reduced agricultural commodity prices and allows farmers to increase their income without threatening food production[35]

SW Deforestation and Second Generation Biofuels

Biomass made from wood is a recent second-generation biofuel that will soon be commercially viable. As long as this is practiced in a sustainable way and ensures that dangers of deforestations are avoided, then it could present a promising alternative to biofuels grown on agricultural land. That way, effects of indirect land-use change (ILUC) will be mitigated and food security will be enhanced [36, 37].

References

1. Groenendijk L (2002) SWOT analysis: strengths, weaknesses, opportunities, threats social science division/ITC/Enschede/ Netherlands. http://pgis-tk.cta.int/m05/docs/M05U04_handout_SWOT_Analysis_Groenendijk_2002.pdf. Accessed 22 Dec 2016
2. Campbell MN (2008) Biodiesel: algae as a renewable source for liquid fuel. Guelph Eng J 1:2–7. http://citeseerx.ist.psu.edu/viewdoc/download?doi=10.1.1.477.2410&rep=rep1&type=pdf. Accessed 21Jan 2017
3. Steer A, Hanson C (2015) Biofuels are not a green alternative to fossil fuels. The Guardian. http://www.theguardian.com/environment/2015/jan/29/biofuels-are-not-the-green-alternative-to-fossil-fuels-they-are-sold-as. Accessed 02 Dec 2016
4. Lane J (2012) 6 Methods to avoid the food v Fuel Dilemma- Sep 05, 2012-BIOFUELS DIGEST. http://oilprice.com/Energy/Energy-General/6-Methods-to-Avoid-the-Food-v-Fuel-Dilemma.html. Accessed 27 Dec 2016
5. Blanco-Canqui H (2016) Growing dedicated energy crops on marginal lands and ecosystem services. Soil Sci Soc Am J 80:845–858. https://doi.org/10.2136/sssaj2016.03.0080. https://dl.sciencesocieties.org/publications/sssaj/abstracts/80/4/845. Accessed 27 Dec 2016
6. Prakash A (2007) Grains for food and fuel—at what price? Inter governmental Group on Grains and Rice, meeting, Istanbul, Turkey. www.fao.org/tempref/unfao/bodies/ccp/gr-ri07/GRRI07CRS5.pdf. Accessed 28 Feb 2017
7. Richard T (2012) Penn State University, The food versus fuel debate. Biofuels 3(6):635–648. http://www.massey.ac.nz/~ychisti/Biofuels Interview.pdf. Accessed 09 Feb 2017
8. Vinocur B, Altman A (2005) Recent advances in engineering plant tolerance to abiotic stress: achievements and limitations. Curr Opin Biotechnol 16:123–132. https://www.ncbi.nlm.nih.gov/pubmed/15831376. Accessed 21 Jan 2017
9. Raswant V, Hart N, Romano M (2008) Biofuel expansion: challenges, risks and opportunities for rural poor people how the poor can benefit from this emerging opportunity paper prepared for the round table organized during the thirty-first session of IFAD's Governing Council, 14 February 2008. https://www.ifad.org/documents/10180/7eaa5557-b360-4928-b10d-6227ab93e2b8. Accessed 12 Dec 2016
10. Youngs H, Somerville C (2012) Growing better biofuel crops, research is underway to reduce the use of food crops for biofuels by shifting to dedicated energy crops and agricultural residues. The Scientist. http://www.the-scientist.com/?articles.view/articleNo/32264/title/Growing-Better-Biofuel-Crops/. Accessed 27 Dec 2016
11. Littlejohn C, Curran T, Rainer W, Wratten S (2015) Farmland, food, and bioenergy crops need not compete for land. Solutions 6(3):34–48. http://www.thesolutionsjournal.com/node/237359. Accessed 12 Dec 2016
12. Worldwatch Institute (2006) Biofues for transportation: global potential and implications for sustainable agriculture and energy in the 21st century. Worldwatch Institute, WashingtonDC, USA. http://cgse.epfl.ch/webdav/site/cgse/shared/Biofuels/BiofuelsforTransport-WorldWatchInstituteandtheGermanGovernment.pdf. Accessed 21 Jan 2017
13. Fulton L, Howes T, Hardy J (2004) Biofuels for transport: an international perspective. International Energy Agency, Paris, France. http://www.iea.org/textbase/nppdf/free/2004/biofuels2004.pdf. Accessed 21 Jan 2017
14. Hu W et al (1999) Repression of lignin biosynthesis promotes cellulose accumulation and growth in transgenic trees. Nat Biotechnol 17:808–812, 1999. http://www.nature.com/nbt/journal/v17/n8/abs/nbt0899_808.htm. Accessed 19 Jan 2017
15. Fu C et al (2011) Genetic manipulation of lignin reduces recalcitrance and improves ethanol production from switchgrass. PNAS 108:3803–3808, 2011. http://www.pnas.org/content/108/9/3803.abstract. Accessed 19 Jan 2017
16. Gille S et al (2011) O-Acetylation of *Arabidopsis* hemicellulose xyloglucan requires AXY4 or AXY4L, proteins with a TBL and DUF231 domain. Plant Cell 23:4041–4053, 2011. http://www.plantcell.org/content/23/11/4041.long. Accessed 19 Jan 2017

17. Chuck SG et al (2011) Overexpression of the maize *Corngrass1* microRNA prevents flowering, improves digestibility, and increases starch content of switchgrass. PNAS 108:17550–17555, 2011. http://www.pnas.org/content/108/42/abstract. Accessed 19 Jan 2017
18. Stewart A, Cromey M (2011) Identifying disease threats and management practices for bioenergy crops. Curr Opin Environ Sustain 3(1–2):75–80. http://www.sciencedirect.com/science/article/pii/S1877343510001144. Accessed 20 Dec 2016
19. Lewandowski I, Scurlock JMO, Lindvall E, Christou M (2003) The development and current status of perennial rhizomatous grasses as energy crops in the US and Europe. Biomass and Bioenergy 25:335–361. http://www.upbiofuel.com/wp-content/uploads/2012/06/dev-status-of-perennial-energy-crops.pdf. Accessed 21 Jan 2017
20. Koh LP, Ghazoul J (2008) Opportunities, B I O LO G I CA L C O NS E RVAT IO N 1 4 1 (2 0 0 8) 2 4 5 0 –2 4 6 0, Elsevier Ltd. http://www.sciencedirect.com/science/article/pii/S0006320708002954. Accessed 21 Jan 2017
21. Kyriakopoulos G, Kolovos GK, Chalikias SM (2009) Environmental sustainability and financial feasibility evaluation of wood fuel biomass used for a potential replacement of conventional space heating sources. Part II: a combined Greek and the nearby Balkan countries case study, Springer. https://doi.org/10.1007/s12351-009-0034-x, http://www.academia.edu/1478356/Environmental_sustainability_and_financial_feasibility_evaluation_of_woodfuel_biomass_used_for_a_potential_replacement_of_conventional_space_heating_sources._Part_II_A_Combined_Greek_and_the_nearby_Balkan_Countries_Case_Study?auto=download. Accessed 19 Jan 2017
22. The World Health Report (2005). http://www.who.int/whr/2005/en/. Accessed 19 Feb 2017
23. Sticklen M (2006) Plant genetic engineering to improve biomass characteristics for biofuels. Curr Opin Biotechnol 17:315–319. https://msu.edu/~stickle1/Sticklen_Curr%20Opinion_paper.pdf. Accessed 21 Jan 2017
24. Sorda G, Banse M, Kemfert C (2010) An overview of biofuel policies across the world. Energy Policy 38(11):6977–6988. https://doi.org/101016/jenpol201006066, http://www.sciencedirect.com/science/article/pii/S0301421510005434. Accessed 27 Dec 2016
25. Pieprzyk B, Kortluke N (2010) Substitution of biofuels for fossil fuels, era: energy research architecture Costa Rica, German Biofuels Industry Association Verband der Deutschen Biokraftstoffindustrie e.V. (VDB). http://www.era-er.com/uploaded/content/proyecto/1368197392.pdf. Accessed 22 Dec 2016
26. Tilman D, Hill J, Lehman C (2006) Carbon-negative biofuels from low-input high-diversity grassland biomass. Science 314:1598–1600. https://www.ncbi.nlm.nih.gov/pubmed/17158327. Accessed 21 Jan 2017
27. U.S. Congress, Office of Technology Assessment, Potential Environmental Impacts of Bioenergy Crop Production-Background Paper, OTA-BP-E-118 Washington, DC: U.S. Government Printing Office, (1993). https://books.google.gr/books?id=vQxJj92e9KoC&pg=PA34&lpg=PA34&dq=energy+crops+soil+quality&source=bl&ots=KJqdkDgoFf&sig=fDPS9j8JSvn8Swjkc_YnhQ6TA8s&hl=el&sa=X&ved=0CGEQ6AEwB2oVChMIsIeR0qLAyAIVR3EUCh3MhAH2#v=onepage&q=energy%20crops%20soil%20quality&f=false. Accessed 03 Dec2016
28. Louise O (2006) Fresco biomass for food or fuel: is there a dilemma? Universiteit van Amsterdam Member of the Supervisory Board of Rabobank Nederland, The Duisenberg Lecture I Singapore. http://www.globalbioenergy.org/uploads/media/0609_Biomass_for_food_or_fuel.pdf. Accessed 12 Dec 2016
29. 7 alternatives as energy crops for marginal lands, Bioenergy Crops Ltd-International consulting for sustainable development. http://bioenergycrops.com/blog/2013/01/23/7-alternatives-as-energy-crops-for-marginal-lands/. Accessed 04 Mar 2017
30. Wicke B et al (2011) The global technical and economic potential of bioenergy from salt-affected soils. Energy Environ Sci 4:2669–2681, 2011. http://pubs.rsc.org/en/content/articlelanding/2011/EE/c1ee01029h#!divAbstract. Accessed 19 Jan 2017

31. Rahmana MM, Mostafizb BS, Paateroa VJ, Lahdelmaa R (2014) Extension of energy crops on surplus agricultural lands: a potentially viable option in developing countries while fossil fuel reserves are diminishing. Renew Sustain Energy Rev 29:108–119. http://www.sciencedirect.com/science/article/pii/S1364032113006321. Accessed 27 Dec 2016
32. Gyulai I (2008) The Biomass Dilemma, CEEweb for biodiversity and the national society of conservationists—friends of the Earth Hungary. http://www.mtvsz.hu/dynamic/biomassen.pdf. Accessed 22 Dec 2016
33. Searchinger T, Heimlich R (2015) Avoiding bioenergy competition for food crops and land. Working Paper, Installment 9 of creating a sustainable food future. Washington, DC: World Resources Institute. http://www.worldresourcesreport.org.https://www.wri.org/sites/default/files/avoiding_bioenergy_competition_food_crops_land.pdf. Accessed 28 Feb 2017
34. Raboni M, Viotti P, Capodaglio GA (2015) A comprehensive analysis of the current and future role of biofuels for transport in the European Union (EU). Rev Ambient Água 10(1):9–21. ISSN 1980-993X. http://dx.doi.org/10.4136/ambi-agua.1492, http://www.scielo.br/scielo.php?script=sci_arttext&pid=S1980-993X2015000100009. Accessed 22 Dec 2016
35. Zilberman D (2012) University of California, Berkeley, USA, The food versus fuel debate. Biofuels 3(6):635–648. http://www.massey.ac.nz/~ychisti/Biofuels Interview.pdf. Accessed 09 Feb 2017
36. Gao Y, Skutsch M, Masera O, Pacheco P (2011) A global analysis of deforestation due to biofuel development. Working Paper 68, Center for International Forestry Research, 30. http://www.cifor.org/publications/pdf_files/WPapers/WP68Pacheco.pdf. Accessed 22 Nov 2016
37. Havlík P, Schneider UA, Schmid E, Böttcher H, Fritz S, Skalský R, Aoko K, de Cara S, Kindermann G et al (2010) Global land use implications of first and second generation biofuel targets. Energy Policy. https://doi.org/10.1016/j.enpol.2010.03.030, http://www.cifor.org/library/3506/a-global-analysis-of-deforestation-due-to-biofuel-development/. Accessed 28 Feb 2017
38. McLamb E (2011) Fossil fuels versus renewable energy resources, Ecology. http://www.ecology.com/2011/09/06/fossil-fuels-renewable-energy-resources/. Accessed 11 Mar 2017

Chapter 9
Conclusions

Food and energy supply are probably the highest priorities for humanity, more so because food production and energy sourcing have been inimical to the environment in the last century. The debate "food versus fuel" has arisen due to the rapid expansion of biofuel production as a step toward limiting fossil fuel use and mitigating climate change through renewable energy. Given the fact that first-generation biofuels are made from important food crops and have high land and water requirements, it is obvious that this debate expresses actual and significant concerns about the threat that biofuels pose to food security.

As discussed in this paper, the investigation of the two choices, i.e., "growing energy crops for biofuel production or for food" has to incorporate important environmental, social, economic, and sustainability aspects in order to encompass the complex interrelationships between energy, food, land, water, productivity, biodiversity, and the environment. The aim was to describe a methodology that can be applied to identify, describe, and even quantify, if possible, the advantages and disadvantages of each of the two options. Consequently, the multicriteria decision technique that was employed was the SWOT analysis. Apart from outlining the strengths, weaknesses, opportunities, and threats associated with each of the two options, SWOT analysis matrices also facilitated the formulation of strategic choices and alternative methods regarding sustainable energy production and secure food supply for a growing world population. The literature overview and analysis presented is a pooling of existing knowledge from international contexts with respect to bioenergy issues and illustrates synergies and conflicts in the bioenergy/environment debate. The following strengths, weaknesses, opportunities, and threats of the two options "using energy crops for biofuel production or for food" are identified by SWOT analysis (Tables 7.4 and 7.6) and summarized below.

Growing energy crops for biofuel production has considerable environmental benefits. Bioenergy is the only carbon negative form of energy and offers a significant contribution to GHG emissions reduction and therefore to reversing the climate change. It is also an important renewable, that enables diversity in energy supply and can increase independence from fossil fuels, thus empowering oil-dependent

A. Paschalidou et al., *Using Energy Crops for Biofuels or Food: the Choice*,
Green Energy and Technology, https://doi.org/10.1007/978-3-319-63943-7_9

countries and avoiding the volatility of oil prices. Moreover, using energy crops for biofuels can lead to the development of the rural economy by increasing producers' incomes and by leading to the creation of direct and indirect jobs associated with biofuels production. Finally, certain energy crops can reduce soil erosion and contribute to soil remediation when they are grown on contaminated or polluted lands.

In contrast, biofuel production, especially if carried out on a large scale, leads to direct and indirect land-use change. This shift of arable land entails a number of drawbacks, the most important of which is that it competes with food production and therefore can lead to increased food prices and food insecurity, particularly for the poorer parts of the population. Moreover, growing energy crops for biofuels often results in large monocultures which have high input demands and stress natural resources, mainly water, with detrimental effects on soil, water, and air quality. It can also lead to deforestation, thus endangering sensitive ecosystems and biodiversity in general, but also to increased GHG emissions. Consequently, biofuel production relies heavily on land availability, and this terrestrial limit, combined with other competing land uses such as agriculture and forestry, presents a major drawback. At the same time, farmers may be reluctant to grow energy crops for biofuel production due to restricted knowledge or information regarding their cultivation, but also because they are sensitive to climate conditions and pest attacks and prone to fires. Furthermore, energy crops that are genetically modified to increase yields and resist pests can be highly invasive and may result in uncontrolled spreading of GMOs. However, the main obstacle for the extended development of biofuels remains their high cost compared to conventional fuels.

There are significant opportunities in bioenergy production, and the most important is the development of a dynamic economic sector that can produce substantial revenues with high growth potential, especially with next-generation biofuels. Moreover, the world agenda on climate change offers incentives and guidelines that promote bioenergy. Many countries and the European Union especially have issued directives and set targets for increased biofuel share in energy supply. Also, energy crops, particularly the perennial ones, present an excellent opportunity for the utilization of marginal or abandoned lands, increasing productivity without competing with food production. Finally, the accumulated experience and recent increased biofuel production has led to important technology advancements that can reduce production cost to make it economically competitive with fossil fuels.

The predominant threat deriving from biofuel production is the food security risks that it entails. Rapid biofuel expansion has already claimed a part of arable land, and earning prospects may attract even more. As a result, reduced food and feed production and higher food prices may aggravate the prolonged food crisis but also may fail to cater for nutrition needs of an increasing world population. In any case, the first to be affected by this are the poorer parts of society and countries that depend on food imports. In addition, in order to overcome the competition with fossil fuels and be cost-effective, biofuel crops present a higher risk of large monocultures and GMO use, with the ensuing environmental impact and the danger of not conforming to crucial sustainability criteria.

On the other hand, using energy crops for food has appreciable advantages, the most vital one being the effect on food security. Food supply, especially for import-dependent countries and urban citizens in developing parts of the world, will benefit immensely. Moreover, the growing world population and the terrestrial limitations of the planet is a challenge that can only be faced with increased food production. Another major strength is that the increasing need to find new lands for growing plants for biofuel production is significantly reduced, along with the risk of further deforestation and the ensuing environment impact. However, rural areas, where unemployment runs high particularly among young people, will lose an important source of income and economic development. A second drawback is that marginal or abandoned lands cannot be utilized for the cultivation of edible crops.

Opportunities that promote food production but also balance energy demand with food requirements are based on research and development of crops and crop processing technologies. Switching to next-generation biofuels and developing new varieties of non-edible energy crops for biofuel production may well prove to be such solutions. Finally, limiting the use of energy crops for biofuels entails the threat of continued energy dependence on fossil fuels and the risk of failure to promote global priority for reversing the climate change.

Considering the findings of the SWOT analysis, targeted strategies and alternative choices emerge for effective energy planning that does not compromise food security.

An obvious alternative strategy is the employment of a diverse variety of renewables to complement bioenergy. Wind and solar energy can cover part of the demand, thus reducing stress on land and food crops.

Another important and easily implementable alternative strategy is using next-generation biofuels. By this diversification of feedstock, using major food crops like soy and maize for biofuels is avoided. Non-edible crops and perennial grasses are an attractive option along with algae or other kinds of lignocellulosic matter like agricultural, forestry, or municipal wastes. Moreover, next-generation energy crops can be cultivated on marginal or agriculturally degraded lands, and thus may not compete with food production.

Biotechnology can play a determining role in future energy planning by developing higher-yielding biofuel non-edible crops, genetically engineered to be more resistant to pests, diseases, drought, flooding, salt, cold, and abiotic stresses, which would insure a stable supply of feedstock. These crops can be grown on salient lands or generally areas where food production is not feasible. This will have to be supported by a development in process technologies that convert next-generation feedstocks to fuels. As a result, rural communities will retain a chance for employment and financial growth. Also, technological developments that enhance biomass yields and augment plants production will result in less land requirements and increased energy supply.

Finally, considering all the aspects of the analysis, the strategy that stands out as the most effective way to overcome weaknesses and face threats is conforming to sustainability principles. If policy makers and governments alike promote and implement agricultural practices that protect natural resources from intensified production then ensure, apart from energy and food security, the conservation of the environment and the protection of biodiversity will also be ensured. An important step toward this approach is the recent EU Directives regarding biomass production enforcing limitations in the origin of feedstock and increasing the minimum requirements in GHG emissions reduction. Biofuel production, as well as food production, should be governed by sustainability standards covering both direct and indirect impacts on the environment and socioeconomic issues. This way, the negative environmental aspect of both biofuel and food production can be mitigated. Sustainable energy and agriculture planning must also include incentives and strategies for promoting smallholders' interests, protecting the poorer parts of the population and encouraging equal opportunity for men and women alike.

In conclusion, it becomes apparent that making a choice between using energy crops for biofuels or for food is a complex and multidimensional issue. By suggesting alternative of courses of action and non-edible feedstocks, humanity might not need to make that choice at all, since food security is paramount but also depletable energy source like fossil fuels cannot continue to be used. However, the food versus fuel debate symbolizes a much deeper concern: People have been on a trajectory of diminishing resources for so long that the threat of non-reversible damage is imminent. Apart from exploring ways to increase energy supply, it is imperative that energy saving and conservation become incorporated in human behavior. Moreover, methods like first-generation biofuels which were initially developed to mitigate climate but instead resulted in depleting natural resources and threatening food security should be avoided. It is advisable and even necessary to incorporate sustainability criteria in policy making. In order to achieve that, research and technology development will require the participation of scientists in wide-ranging fields, including ecology, environmental sciences, plant biology, microbiology, biochemistry, chemistry, genetics, and economics under the common objective to not deplete natural resources and to ensure the welfare of the planet for future generations.

Appendix

From 2011 to 2030, advanced biofuels have the potential to create 8 million man-years of employment around the world and different types of jobs and potentials for each region. Also, transforming agricultural residues into advanced biofuels could create economic growth, reduction of greenhouse gas emissions, and energy security by 2030 [126].

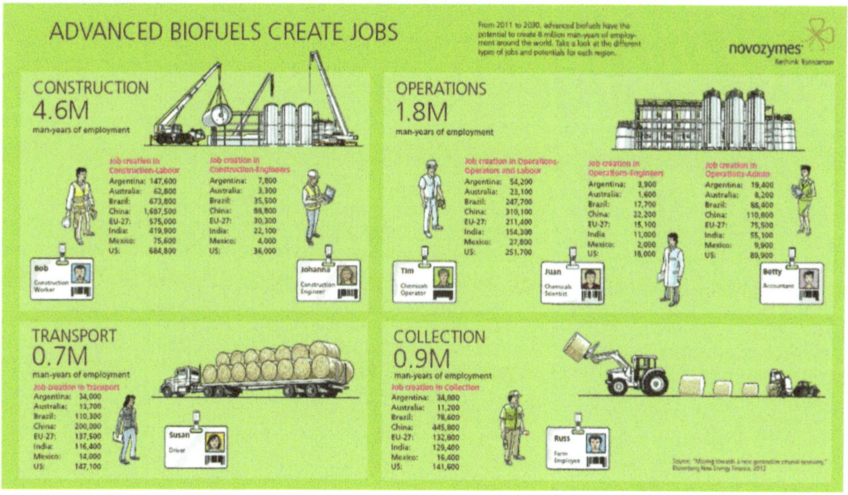

Fig. A.1 Advanced biofuels create jobs (*Source* [126])

A. Paschalidou et al., *Using Energy Crops for Biofuels or Food: the Choice*,
Green Energy and Technology, https://doi.org/10.1007/978-3-319-63943-7

How much land is available for energy farming depends or how much is needed for the production of food and feed depends on agricultural productivity and changing dietary patterns (Table A.1) [95].

Table A.1 Required arable land for ten different energy crops [assuming 10% substitution of current global energy demand (450 EJ/a) or 25% of current oil demand (5000 bn l/a), assuming global arable land is 2.5 bln ha] [95]

Crop	Yield l/ha	Yield GJ/ha	Billion hectares	% of arable land
Barley	1100	19.8	2.27	91
Wheat	2500	45	1.00	40
Corn	3000	54	0.83	33
Sugar beet	5000	90	0.50	20
Sugarcane	5800	104.4	0.43	17
Soybean	500	9	5.00	200
Sunflower	900	16.2	2.78	111
Rapeseed	1100	19.8	2.27	91
Jatropha	1800	32.4	1.39	56
Oil plam	4500	81	0.56	22

Crop yields are based on Fulton et al, IEA 2004 [95]

Since 2004, (Table A.2; Fig. A.1) the share of renewable sources in gross final consumption of energy grew significantly in all Member States of EU [104] (Figs. A.2 and A.3).

Table A.2 Share of energy from renewable sources (in % of gross final energy consumption) (*Source* [104])

a/a		2004	2011	2012	2013	2014	2020 target
1.	**EU**	**8.5**	**13.1**	**14.3**	**15.0**	**16.0**	**20**
2.	Belgium	1.9	6.2	7.2	7.5	8.0	13
3.	Bulgaria	9.4	14.3	16.0	19.0	18.0	16
4.	Czech Republic	5.9	9.5	11.4	12.4	13.4	13
5.	Denmark	14.9	23.5	25.6	27.3	29.2	30
6.	Germany	5.8	11.4	12.1	12.4	13.8	18
7.	Estonia	18.4	25.5	25.8	25.6	26.5	25
8.	Ireland	2.4	6.6	27.1	7.7	8.6	16
9.	**Greece**	**6.9**	**10.9**	**13.4**	**15.0**	**15.3**	**18**
10.	Spain	8.3	13.2	14.3	15.3	16.2	20
11.	France	9.4	11.1	13.4	14.0	14.3	23
12.	Croatia	23.5	25.4	26.8	28.1	27.9	20
13.	Italy	6.3	12.9	15.4	16.7	17.1	17
14.	Cyprus	3.1	6.0	6.8	8.1	9.0	13
15.	Latvia	32.8	33.5	35.7	37.1	38.7	40
16.	Lithuania	17.2	20.2	21.7	23.0	23.9	23
17.	Luxembourg	0.9	2.9	3.1	3.6	4.5	11
18.	Hungary	4.4	9.1	9.6	9.5	9.5	13
19.	Malta	0.1	1.9	2.9	3.7	4.7	10
20.	Netherlands	2.1	4.5	4.7	4.8	5.5	14
21.	Austria	23.3	30.8	31.6	32.3	33.1 3	34
22.	Poland	6.9	10.3	10.9	11.3	611.4	15
23.	Portugal	19.2	24.7	25.0	25.7	27.0	31
24.	Romania	17.0	21.4	22.8	23.9	24.9	24
25.	Slovenia	16.1	20.2	20.9	22.5	21.9	25
26.	Slovakia	6.4	10.3	10.4	10.1	11.6	14
27.	Finland	29.2	32.8	34.4	36.7	38.7	38
28.	Sweden	38.7	49.0	51.1	52.0	52.6	49
29.	United Kingdom	1.2	4.2	4.6	5.6	7.0	15
30.	Iceland	58.9	71.6	73.2	72.2	77.1	64
31.	Norway	58.1	64.8	65.9	66.7	69.2	67.5

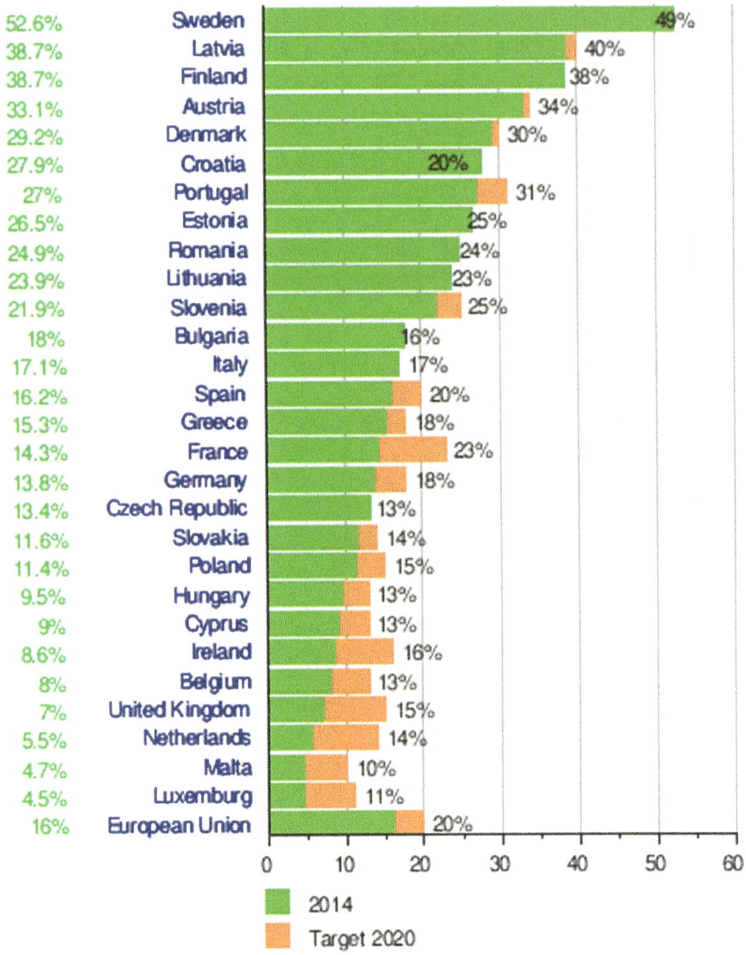

Fig. A.2 Share of energy from renewable sources (in % of gross final energy consumption) (*Source* [104])

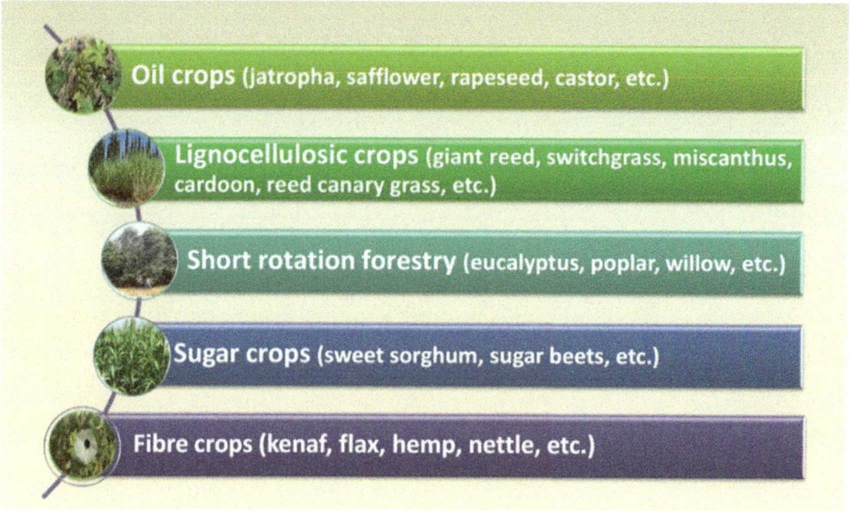

Fig. A.3 High biomass yielding non-food crops (*Source* [173])

The manufacturer's authorised representative in the EU is Springer Nature Customer Service Centre GmbH, Europaplatz 3, 69115 Heidelberg, Germany. If you have any concerns regarding our products, please contact ProductSafety@springernature.com

Printed and bound by CPI Group (UK) Ltd, Croydon, CR0 4YY
30/09/2026
02225664-0008